Geomorphic Setting, Aquatic Habitat, and Water-Quality Conditions of the Molalla River, Oregon, 2009–10

By Kurt D. Carpenter, Christiana R. Czuba, Christopher S. Magirl, Mathieu D. Marineau, Steve Sobieszczyk, Jonathan A. Czuba, and Mackenzie K. Keith

Prepared in cooperation with the Molalla River Improvement District

Scientific Investigations Report 2012–5017

U.S. Department of the Interior
U.S. Geological Survey

U.S. Department of the Interior
KEN SALAZAR, Secretary

U.S. Geological Survey
Marcia K. McNutt, Director

U.S. Geological Survey, Reston, Virginia: 2012

For more information on the USGS—the Federal source for science about the Earth, its natural and living resources, natural hazards, and the environment, visit http://www.usgs.gov or call 1–888–ASK–USGS.

For an overview of USGS information products, including maps, imagery, and publications, visit http://www.usgs.gov/pubprod

To order this and other USGS information products, visit http://store.usgs.gov

Suggested citation:
Carpenter, K.D., Czuba, C.R., Magirl, C.S., Marineau, M.D., Sobieszcyk, S., Czuba, J.A., and Keith, M.K., 2012, Geomorphic setting, aquatic habitat, and water-quality conditions of the Molalla River, Oregon, 2009–10: U.S. Geological Survey Scientific Investigations Report 2012–5017, 78 p.

Contents

Executive Summary 1

Introduction 3

    Purpose and Scope 5

    River Centerline and Geomorphic Flood-Plain Convention 5

    Study Area and Sampling Locations 7

Description of Molalla River Basin 11

    Geology...11

    Hydrology 13

    Designated Beneficial Uses and Status of Fish Populations 14

    Aquatic Habitat and Water Quality Conditions..16

    Land Use and Anthropogenic Impacts..17

Geomorphology 18

    Geomorphic Reaches...18

        Channel Characterization within Geomorphic Reaches 19

        Flood-Plain Morphology 22

    Key Longitudinal Trends along the River Corridor 25

        Water-Surface Elevation 25

        Water Depth and Residual Depths..27

        River-Profile Convexities 28

        Particle Size 29

        Cross-Section Data 30

    Analysis of Stage-Discharge Relation at Gaging Station 31

    Spatial Analysis of Flood Plain 31

        Methodology of Digitization 32

        Quantitative Trends...33

        Qualitative Trends after the 1930s 37

        Channel Response to Flooding 43

Synthesis of Geomorphic Analyses..43

Streamflow, Water Quality, and Algal Conditions 44

    Field Parameters..45

    Nutrients..50

    Benthic Community Conditions—Algae and Macroinvertebrates.................................55

Synthesis of Water-Quality and Benthic Community Analyses 63

Geomorphic and Water-Quality Factors Affecting Algae and River Food Webs 63

    Multivariate Analyses of Diatom Assemblages..64

Aquatic Habitat and Water-Quality Conditions for Fish..68

Implications for Resource Management 69

Potential Future Studies..70

Acknowledgments 71

References Cited...71

Appendixes and Evaluation of Quality Assurance Data 77

Contents—Continued

Appendix A. Particle size distributions collected using Wolman (1954) Pebble Counts on nine freshly deposited bars along the river corridor of the Molalla River, Oregon, September–October 2010 77

Appendix B. Periphyton species abundance data (cell density) in the Molalla River, Oregon, August–September 2010 77

Appendix C. Periphyton species abundance data (algal biovolume) in the Molalla River, Oregon, August–September 2010 77

Appendix D. Quality assurance data for total and dissolved nutrients in the Molalla River, Oregon, August–September 2010 77

Appendix E. Quality assurance data for replicate periphyton species composition samples from the Molalla River, Oregon, 2010 77

Appendix F. Nutrient and field parameter data for the Molalla River, Oregon, June 26, 2000 77

Appendix G. Field parameter data for the Molalla River, Oregon, 2000 77

Figures

Figure 1. Map showing location and extent of the Molalla River study reach within the Molalla-Pudding River basin, Oregon 4

Figure 2. Map showing geomorphic flood plain, river centerline and river kilometer stationing (Rkm), geomorphic flood-plain reaches, and geomorphic flood-plain centerline and flood-plain kilometer transect stationing (FPkm), Molalla River, Oregon 6

Figure 3. Map showing locations of point-specific data collection sites along the Molalla River, Oregon 8

Figure 4. Graph showing monthly mean precipitation and discharge of the Molalla River measured at the U.S. Geological Survey streamflow-gaging station Molalla River near Canby, Oregon 13

Figure 5. Graph showing enhanced annual peak streamflow for the Molalla River near Canby, Oregon, 1909–2009 15

Figure 6. Graph showing geomorphic flood-plain width along the Molalla River, Oregon ... 19

Figure 7. Photographs of typical channel characteristics for each geomorphic reach in the lower Molalla River, Oregon 21

Figure 8. Maps showing height above water surface of the Molalla River centered on each geomorphic reach 23

Figure 9. Graphs showing Light Detection and Ranging (LiDAR)-derived longitudinal water-surface elevation profile of the Molalla River, Oregon, with elevation plotted in normal coordinates and semi-log coordinates 26

Figure 10. Graph showing water depth data from the Molalla River, Oregon, as measured in July 2010 by a fathometer mounted to boat floating down the river thalweg 27

Figure 11. Graph showing deviation of water-surface elevation profile from second-order polynomial trendline, and measured channel-bed and bedrock within the channel, Molalla River, Oregon 28

Figure 12. Boxplot showing distribution of particle-size data collected using Wolman (1954) pebble counts on nine freshly deposited bars along the river corridor of the Molalla River, Oregon 29

Figures—Continued

Figure 13. Diagram showing comparison of measured geometry of five river cross sections surveyed on the Molalla River, Oregon 30

Figure 14. Graph showing trends in stage of water-surface elevation at the U.S. Geological Survey streamflow-gaging station 14200000, Molalla River near Canby, Oregon 31

Figure 15. Graph showing historical changes in channel sinuosity, averaged over geomorphic flood-plain reaches, for the Molalla River, Oregon, 1994–2009 34

Figure 16. Graph showing historical changes in active-channel width, averaged over each 1-FPkm subreach segment, for the Molalla River, Oregon, 1994–2009 34

Figure 17. Graph showing average annual channel-migration rates, averaged over each 1-FPkm subreach segment, for three time periods between aerial imagery, Molalla River, Oregon, 1994–2009 35

Figure 18. Graph showing percentage of channel banks confined by bedrock, revetment, or Missoula Flood deposits along the Molalla River, Oregon, in 2010 35

Figure 19. Graphs showing historical changes in specific vegetated bar area averaged over each 1-FPkm subreach segment, separated by bar type and vegetation density, along the Molalla River, Oregon, 1994–2009 36

Figure 20. Paneled mosaics of imagery of the river corridor in geomorphic reach 3 for all available sets of aerial imagery 38

Figure 21. Aerial imagery of the Molalla River in geomorphic reach 2 near Canby, Oregon, in 1936 and 2009, showing the relative stability of channel form and location during most of the 20th century 41

Figure 22. Aerial imagery of the Molalla River in geomorphic reach 1 from near the confluence of the Willamette River in 1936 and 2009, showing a wide and active flood plain between the Pudding River and the Molalla River, Oregon 42

Figure 23. Aerial imagery of the Molalla River, Oregon, in geomorphic reach 4 just upstream of Highway 213 bridge to about FPkm 23 in 1948 and 2009 42

Figure 24. Graph showing afternoon water temperature and specific conductance values for August–September 2000, August–September 2010, and longitudinal streamflow for various periods in the Molalla River, Oregon 48

Figure 25. Graph showing water temperatures in the Molalla River, Oregon, August 2001 49

Figure 26. Graphs showing concentrations of dissolved oxygen and pH levels in the Molalla River, Oregon in August 2000, September 2000, August 2010, and September 2010 50

Figure 27. Graph showing long-term trend in pH values in the lower Molalla River, 1965–2010 50

Figure 28. Graphs showing longitudinal nutrient concentrations in the main-stem Molalla River, Oregon, August and September 2010 and June 2000 51

Figure 29. Graph showing summertime dissolved inorganic nitrogen and phosphorus concentrations in the Molalla River, Oregon, 1990–2010 55

Figure 30. Graph showing longitudinal pattern in periphyton biomass (chlorophyll-*a*) in the main-stem Molalla River, Oregon, July and August 2010 56

Figure 31. Graphs showing longitudinal pattern in diatom indicators of water-quality conditions in the Molalla River, Oregon, in July and August 2010 61

Figure 32. Multidimensional scaling (MDS) ordination plot of algal samples from the Molalla River, Oregon, for the combined July and August 2010 samples 64

Tables

Table 1. Conversion between river-centerline stationing, U.S. Geological Survey (USGS) topographic map river-mile stationing, and geomorphic flood-plain stationing for the Molalla River, Oregon 7

Table 2. Sites sampled on the Molalla River, Oregon, and details of specific data collected during July–October 2010 9

Table 3. Summary of data used to create an enhanced peak discharge record for the Molalla River, Oregon 14

Table 4. Discharge values of given annual exceedance probabilities for peak flows for the Molalla River using the U.S. Geological Survey gaging record from Canby, weighted regression estimates, and the enhanced record 15

Table 5. Upstream and downstream extents of the six geomorphic reaches along the lower Molalla River, Oregon 19

Table 6. Residual pool depth summary statistics by geomorphic reach in the Molalla River, Oregon, for residual pools greater than 1.0 meter 28

Table 7. Aerial imagery used in spatial analysis of the Molalla River, Oregon, flood plain ... 32

Table 8. Diel field parameter data and diel ranges in the main-stem Molalla River, Oregon, August–September 2010 46

Table 9. Nutrient concentration data for the main-stem Molalla River, Oregon, 2010 52

Table 10. Nutrient concentrations for reference streams in the Cascade Range and Williamette Valley, Oregon, and nutrient and algal criteria to prevent nuisance conditions in streams 53

Table 11. Periphyton biomass and qualitative descriptions of benthic communities in the main-stem Molalla River, Oregon, July–August 2010 58

Table 12. Autecological preferences and indicator qualities of algal taxa identified in the Molalla River, Oregon, July and August 2010 59

Table 13. Diatom indicator species (guilds) in the Molalla River, Oregon, July and August 2010 60

Table 14. Physical characteristics for five sites sampled for water quality and algae in the Molalla River, Oregon 66

Table 15. Summary of BEST variables and top models explaining patterns in diatom species relative abundance in the Molalla River, Oregon, July and August 2010 67

Conversion Factors and Datums

Conversion Factors

Multiply	By	To obtain
Length		
centimeter (cm)	0.3937	inch (in.)
millimeter (mm)	0.03937	inch (in.)
meter (m)	3.281	foot (ft)
kilometer (km)	0.6214	mile (mi)
Area		
square meter (m^2)	0.0002471	acre
square kilometer (km^2)	0.3861	square mile (mi^2)
Volume		
liter (L)	33.82	ounce, fluid (fl. oz)
liter (L)	1.057	quart (qt)
liter (L)	0.2642	gallon (gal)
cubic meter (m^3)	264.2	gallon (gal)
cubic centimeter (cm^3)	0.06102	cubic inch (in^3)
liter (L)	61.02	cubic inch (in^3)
cubic meter (m^3)	35.31	cubic foot (ft^3)
Flow rate		
meter per second (m/s)	3.281	foot per second (ft/s)
cubic meter per second (m^3/s)	35.31	cubic foot per second (ft^3/s)
Mass		
microgram (µg)	1,000	milligram (mg)
gram (g)	0.03527	ounce, avoirdupois (oz)
kilogram (kg)	2.205	pound avoirdupois (lb)
megagram (Mg)	1.102	ton, short (2,000 lb)
megagram (Mg)	0.9842	ton, long (2,240 lb)
Hydraulic gradient		
meter per meter (m/m)	3.281	foot per foot (ft/ft)
meter per kilometer (m/km)	5.27983	foot per mile (ft/mi)

Temperature in degrees Celsius (°C) may be converted to degrees Fahrenheit (°F) as follows:

$$°F = (1.8 \times °C) + 32.$$

Specific conductance is given in microsiemens per centimeter at 25 degrees Celsius (µS/cm at 25°C).

Concentrations of chemical constituents in water are given either in milligrams per liter (mg/L) or micrograms per liter (µg/L).

Algal biomass values are given in milligrams per square meter (mg/m^2).

Algal cell density values are given in numbers per square centimeter (#/cm^2).

Algal biovolume values are given in cubic micrometers per square centimeter (µ^3/cm^2).

Conversion Factors and Datums—Continued

Datums

Vertical coordinate information is referenced to the North American Vertical Datum of 1988 (NAVD 88).

Horizontal coordinate information is referenced to the North American Datum of 1983 (NAD 83).

Elevation, as used in this report, refers to distance above the vertical datum.

Geomorphic Setting, Aquatic Habitat, and Water-Quality Conditions of the Molalla River, Oregon, 2009–10

By Kurt D. Carpenter, Christiana R. Czuba, Christopher S. Magirl, Mathieu D. Marineau, Steve Sobieszczyk, Jonathan A. Czuba, and Mackenzie K. Keith

Executive Summary

This report presents results from a 2009–10 assessment of the lower half of the Molalla River. The report describes the geomorphic setting and processes governing the physical layout of the river channel and evaluates changes in river geometry over the past several decades using analyses of aerial imagery and other quantitative techniques.

The peak-flow hydrology in the Molalla River has been characterized by a series of large floods during the 1960s and 1970s, a period of relatively small peak flows from 1975 to 1995, and a relative increase in severity of events in the past 15 years. Although incomplete, the gaging record for the early 20th century showed only modest high flows. The flood chronology since 1960 has affected the geomorphology of the river corridor, principally by increasing the active-channel width. The area affected by channel migration in the late 20th century, however, was reduced by the construction of revetments along the river corridor which acted to contain channel movement.

The study area along the Molalla River was divided into six unique geomorphic reaches. The upper-most reach, designated GR6, is a narrow, bedrock-controlled reach with ample shade and large riffles. The next downstream reach, GR5, is also largely bedrock controlled but has a wider flood plain and active channel-migration zone. The longest geomorphic reach, GR4, has a wide channel-migration zone with many strategically placed revetments that work in concert with bounding bedrock to the northeast to suppress overall channel movement. In contrast, GR3 is a wide, active reach that responds more dramatically to flood and non-flood periods than the other geomorphic reaches. The anthropogenically confined GR2, adjacent the City of Canby, has relatively little historical channel movement and relatively few gravel bars. Finally, the farthest downstream reach, GR1, is an actively meandering reach that most closely resembles its pre-development state.

Detailed analysis of aerial imagery from 1994, 2000, 2005, and 2009 showed that channel-migration activity and active-channel widths were greater in GR3 than in any other geomorphic reach and were related directly to the timing and magnitude of high flows. Similarly, the revegetation of exposed bars is significant in GR3 and elsewhere when large floods do not occur. A qualitative analysis of older aerial imagery dating back to 1936 showed that the recent channel-migration activity in GR3 is no greater than it was historically. Channel-migration activity in GR2, GR4, and GR5 was reduced relative to historical rates as a consequence of the construction of revetments and encroachment along the river corridor.

Analyses of the longitudinal water-surface profile first suggested a possible accumulation of alluvium in GR3, but subsequent analysis of the shape of the longitudinal profile juxtaposed against bedrock outcrops in the river channel showed that the river is largely flowing over a shelf of bedrock and not filling with sediment.

Water-quality, benthic algae, and benthic invertebrate conditions were examined during summer low-flow periods to determine the overall health of the river and to provide possible insights into the physical or chemical influences on diatom assemblages.

A wetter than normal spring in 2010 resulted in higher-than-normal flows in July and August that may have delayed the algal growing season and limited the accrual of algal biomass in the river. Longitudinal changes in water quality, including downstream increases in water temperature and specific conductance, were observed in the Molalla River during August and September. Such patterns are typical of many rivers receiving inputs from anthropogenic sources in the flood plain, including agricultural and rural residential lands (Milk and Gribble Creek basins) as well as some urban runoff in the lower river.

Nutrient concentrations in the Molalla River were generally low at most sampling sites but did increase at the Goods Bridge and Knights Bridge sites, presumably from a greater influence from anthropogenic sources that enter the river from tributaries, agricultural irrigation returns, or groundwater in the lower basin. Nitrate concentrations at Glen Avon and Knights Bridges exceeded their respective reference values for streams in the Cascade Range and Willamette Valley. Although the nitrate-nitrogen concentrations were somewhat elevated, phosphorus, in contrast, is relatively much less abundant in the Molalla River. N:P ratios for soluble, biologically available nitrogen and phosphorus were lower in the upper middle reaches (less than 5), but the absolute concentrations of orthophosphorus (0.010 milligrams per liter or less in July) suggest that attached periphytic algae in the river may be limited by phosphorus concentrations or some other factor, but probably not by nitrogen. The Molalla River has lower phosphorus concentrations than other rivers draining the Cascade Range because the phosphate-rich rocks of the Oregon High Cascades, prevalent in other drainages, are not present in the Molalla River basin, which is wholly contained within the Western Cascade Range geologic province.

The 2010 algal growing season was delayed due to an unusually cold and wet spring, which produced streamflows 12–18 percent higher than normal in July and August and could have limited the accrual of periphyton biomass in the river. Nevertheless, a healthy biofilm of diatoms and other types of algae developed in the shallow riffle habitats during July, covering the entire stream channel in some areas. Generally, riffle habitats appeared healthy, with little sediment and low substrate embeddedness (that is, the degree of infilling of fine sediments around gravels and cobbles) was less than 5 percent at all sites except the Knights Bridge site, where embeddedness was about 10 to 25 percent higher.

Algal biomass levels in July were moderate, ranging from 30 to 55 mg of chlorophyll-a per square meter, and the high densities of benthic macroinvertebrate grazers in the riffles suggests that the accumulation of algae (biomass levels) may have been limited by these herbivores. In August, however, a benthic bloom of filamentous green algae (*Cladophora glomerata*) increased algal biomass in the lower river, with nuisance levels at the Knights Bridge site. Higher nutrient concentrations (both nitrate and orthophosphate) combined with fewer invertebrate grazers (mostly snails) likely contributed to the higher biomass at this site. Long filaments of *Cladophora* also were observed in the area near the Canby drinking-water treatment plant, where in previous years, algae have clogged water intakes during periods of senescence when algae detach from the river bed and enter the intake. In 2010, algal biomass conditions were not as severe and the intakes were not affected.

Distinct fluctuations in concentrations of dissolved oxygen and in pH levels from algal photosynthesis were observed at all sites sampled, with the largest diel changes and highest daily maximum values occurring at the two most downstream sites, particularly at Knights Bridge. Although some relatively high pH values were measured (as much as 8.4 units), none of the pH measurements exceeded State of Oregon water-quality standards, even in the afternoon hours on warm sunny days. Dissolved oxygen concentrations at Goods Bridge and Knights Bridge did not meet the 8 milligrams per liter criteria in the early morning hours, but compliance with the standards is only evaluated with 30-day average minimum values, which were not available. Relative to the salmon spawning criteria, for which the data collected during this study applies only to the Glen Avon Bridge site in September, water temperature, pH, and concentrations of dissolved oxygen all met the state standard in effect.

Thirty-three species of algae were identified in the Molalla River, including fast growing small diatoms and very large stalked diatoms, filamentous green and blue-greens, and a few planktonic forms of green and blue-green algae that may have washed into the river from an upstream pond. The occurrence of high-biomass forming types of algae in the river, including filamentous greens such as *Cladophora* and large stalked diatoms such as *Cymbella* and *Gomphoneis*, could be a concern for fish populations because of the potential for smothering fish redds or by impacting benthic invertebrate populations that feed fish.

Together, most of these algae (and overall algal biomass) are typical of generally high quality waters with little organic pollution, high concentrations of dissolved oxygen, and alkaline pH. The relatively high percentage of eutrophic taxa does, however, suggest some degree of nutrient enrichment in the river, despite the relatively low concentrations observed at most sites. Uptake of dissolved nutrients by algae, and inputs of additional nutrients, complicates interpretations regarding nutrient concentrations in the river, especially because samples were collected during the summer growing season.

Although the bulk of the diatom species generally were similar among at least the four upstream sampling sites, the multivariate ordination suggests a downstream trend in assemblage structure from the Glen Avon Bridge site to the Highway 213 Bridge. The next downstream site, at Goods Bridge, near the downstream end of the alluvial GR3 reach, however, plotted closer to the most upstream site at Glen Avon Bridge, which indicates a change in assemblage structure. The algal indicator species analysis showed a change in species composition at the Goods Bridge site, including decreases in eutrophic diatoms, increases in the relative abundance of oligotrophic diatoms, and an increase in diatoms sensitive to organic pollution that suggests an improvement in water quality conditions. Although this may be related to the enhanced water exchange into and out of the streambed in the alluvial reach, and such hyporheic activity could work to clean the river of organic compounds and nutrients, small decreases in water quality (lower concentration of dissolved oxygen, and higher conductance and nutrient concentrations) were observed between the Highway 213 and Goods Bridge sites.

The multivariate analysis relating the diatom species composition data to the geomorphic and water-quality variables indicated that the presence of local gravel bars, bedrock, exposure to the sun (open canopy), and pH had a significant role in shaping the diatom assemblage structure. Although there was a high percentage of similarity among samples, many of these factors have the potential to affect diatoms and other algae through various interrelated mechanisms that relate to channel mobility and associated effects on light available for algal photosynthesis, for example, and other potential factors.

Although only qualitatively addressed for this study, benthic macroinvertebrates, including mayflies, caddisflies, and stoneflies, were abundant in the Molalla River and indicate a high degree of secondary production in the riffles throughout the study reach. Snails, another voracious grazer of algae, also were relatively abundant at the Goods Bridge and Knights Bridge sites. Additionally, large numbers of the large caddisfly larvae *Dicosmoecus* were observed throughout most of the lower river in a range of depths and habitats. The large densities of these grazers, combined with the moderate level of algal biomass, suggest that invertebrate grazers could have limited the accrual of algae during summer 2010, an assertion that could be evaluated with further study. In northern California's Eel River, high abundances of *Dicosmoecus* were detected in summers following winters that lacked bankfull flow, as was the case for the Molalla River in water year 2010. The lack of disturbance might explain the high abundance of these herbivores in the Molalla River.

The information from this study can be used to adapt management strategies for the Molalla River and its flood plain. These strategies may assist in developing and maintaining a healthy river environment that includes high-quality water for aquatic life and human consumption.

Introduction

The Molalla River in northwestern Oregon is a tributary of the Willamette River, joining the Willamette River about 15 kilometers (km) upstream of Oregon City (fig. 1). The Molalla River is valued for many attributes, including its runs of winter steelhead and spring Chinook salmon, high-quality drinking water, and, in summer, refreshing swimming holes. The Molalla River basin also is home to numerous species of wildlife, including more than 80 species of mammals and 150 species of birds (Alan Gallagher, Molalla River Improvement District, written commun., 2011). The upper Molalla River starts in the Cascade Range foothills, in the Table Rock National Wilderness, in an area surrounded by national forests and private forestland. From vantage points such as U.S. Forest Service (USFS) watchtowers and natural mesas, one can view the surrounding countryside of green forests intersected by deep canyons. The recent Congressional nomination of the upper 35 km of the Molalla River and the Table Rock Fork Molalla River for Wild and Scenic River status is testament to the river's uniqueness with spectacular views, interesting geology, and opportunities for camping, hiking, trout and salmon fishing, wildlife viewing, and whitewater kayaking (Heyn and Bassett, 2009).

Located between the Clackamas River to the north and the Pudding River to the southwest, the 79-km long Molalla River drains a 900-km² basin at elevations from 21 m to about 1,510 m (4,940 ft). Stream slope averages 0.0120 m/m in the upper basin and 0.0025 m/m in the lower

basin (Bureau of Land Management and U.S. Forest Service, 1999). Nevertheless, with its headwaters located in the rain-snow transition area, and as one of the few free-flowing unregulated rivers in the Pacific Northwest, the Molalla River is susceptible to large rain and rain-on-snow events that produce large peak-flow discharges. Although peak flows are commonly important in forming and maintaining fish habitat, these large events can also cause landslides in the catchment, deliver sediment to the river, erode banks, scour fish spawning areas, and damage roads and homes. Areas just to the south of Canby that are susceptible to flooding include South Alder Creek Lane, South Vale Garden Road, the housing project on South Elm Street, and certain areas near Highway 99E and Canby Grove (Alan Gallagher, Molalla River Improvement District, written commun., 2011).

Observations of gravel bars, infilling of meanders, and channel scars filled with sediment indicate that bedload transport by the Molalla River can be substantial, but it is unclear to what degree sediment input from upstream land-use activities may influence sedimentation in the main-stem river and potential reductions in flood-conveyance capacity of the channel. No comprehensive analysis has been completed that documents the geomorphic setting of the Molalla River or describes the historical and contemporary trends in river-channel position or character over time. Cole (2002) reported on a few aspects of the geomorphic nature of the river at six locations extending up the Molalla River to the Table Rock Fork Molalla River and found downward trends (in a downstream direction) in percentages of large substrate, erosional habitat, and percent canopy cover, and upward trends in fine-grained sediment in the channel and percent substrate embeddedness (degree of sediment infilling around riffle cobbles). Later, Cole and others (2004) reported on channel confinement, gradient, stream substrate, widths, and depths in the main-stem Molalla River. The geomorphic character of the river has potential to adversely affect aquatic life through, for example, channel widening or other processes that deliver sediment, so understanding the dependencies between the health of the aquatic life and the physical setting is needed for integrated and adaptive management of the river as an ecological resource.

One of the information needs identified by Cole and others (2004) was a field-based survey of the channel geomorphology and channel habitat types in the main-stem river that would collect baseline data on existing conditions. Additionally, concerns about the health of fish populations and declining fish runs have generated interest in documenting the current status of the aquatic habitat available for fish and water-quality conditions in the river. To address these issues, the Molalla River Improvement District, Molalla River Watch, Oregon Department of Fish and Wildlife, and others requested that the U.S. Geological Survey (USGS) conduct a geomorphic and aquatic habitat survey in the lower river.

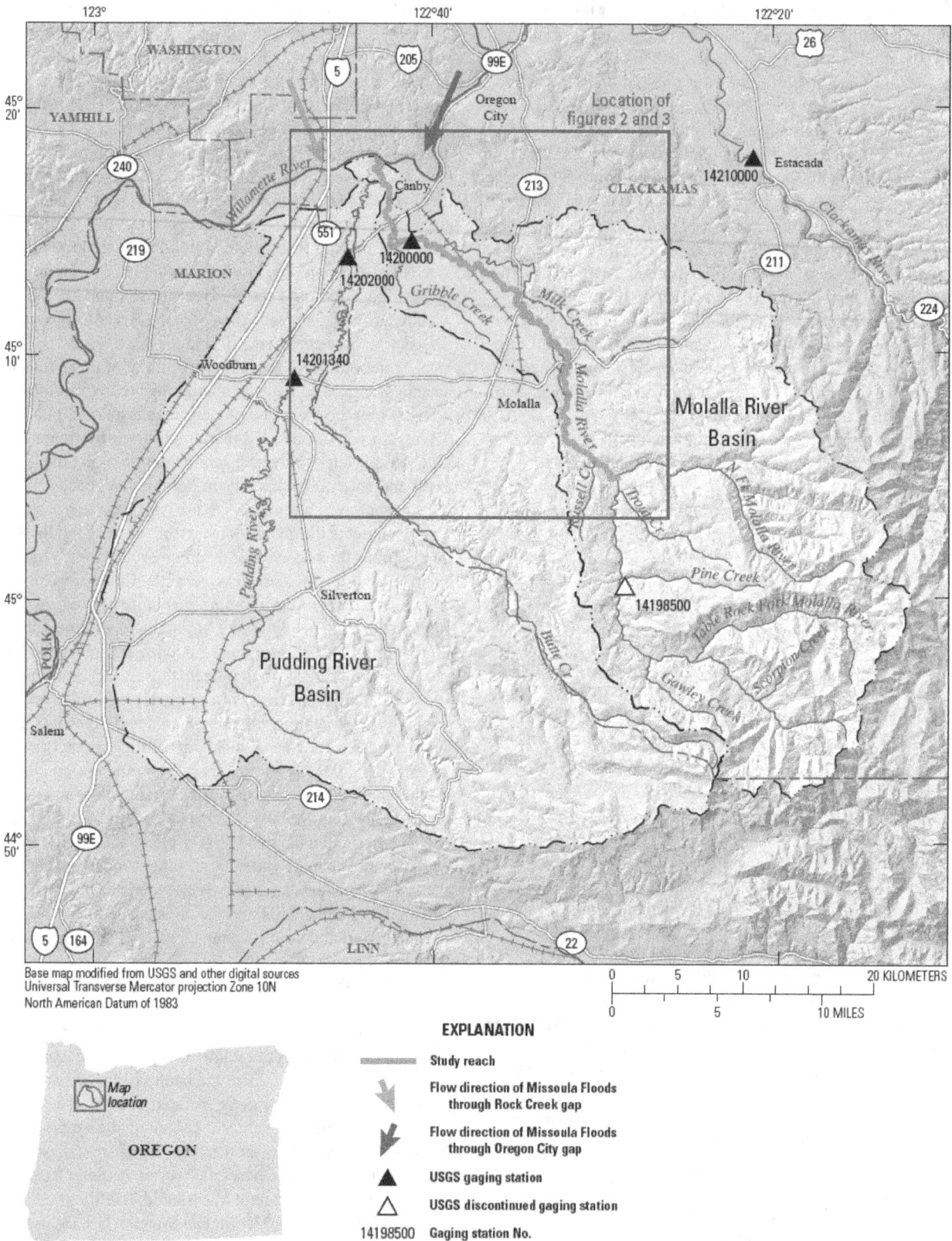

Figure 1.   Location and extent of the Molalla River study reach within the Molalla-Pudding River basin, Oregon.

The objectives of this study were to (1) complete a geomorphic and aquatic habitat characterization of the lower Molalla River to understand the factors driving current conditions in the river; (2) characterize the water quality, benthic algae, and invertebrate conditions; and (3) evaluate potential interactions between algal assemblages and the geomorphic and water-quality parameters. In addition, given ongoing concerns about potential nutrient enrichment, bacteria, proliferations of nuisance algae in the river, and impacts on water quality such as low levels of dissolved oxygen (DO) and high pH, this study also included two surveys of algal conditions to document the current biomass levels and species composition in the lower river during summer. Because of the often strong control that benthic invertebrate grazing can have on algal populations and their importance as a food resource for fish, qualitative surveys for benthic invertebrates were also conducted during algal sampling.

Purpose and Scope

In this report, the results of the analyses of geomorphic conditions and aquatic habitat of the lower Molalla River are presented, including the current conditions and a review of available data and historical photographs to assess changes in the channel width and position over time. This information is used to identify the primary geomorphic controls on channel form. The geomorphic data were combined with the water-quality data to determine if a correlation existed between physical and biological factors and conditions in the river system. Interpretation from this study can be used to complement data and information collected previously by others to inform future research, develop action plans to guide stream-restoration activities, lead management strategies to improve aquatic habitat and water-quality conditions, and possibly reduce the frequency or severity of flooding.

River Centerline and Geomorphic Flood-Plain Convention

Analyses in the study were facilitated by the use of two coordinate systems: (1) a river centerline that follows the low-flow channel water surface, thereby enabling description of variables along the primary river flow path and (2) a stationing system based on the geomorphic flood plain that is invariant over the historical time frame and allows comparison of variables independent of the location of the active channel. When describing the right or left bank of the river or the flood plain, the perspective of the observer is as if looking downstream.

Longitudinal location, or stationing, along the river using the river-centerline coordinate system is expressed as a river kilometer (Rkm) distance upstream of the confluence of the Molalla River with the Willamette River (fig. 2; table 1). Because one goal of delineating the river centerline was to generate an approximate longitudinal water-surface elevation profile throughout the study reach, the river centerline was drawn along the path of minimum elevation of the river channel in a digital elevation model generated with airborne light detection and ranging (LiDAR) data. This path was assumed to best represent the flow path of the river at the time of the LiDAR flights. The LiDAR data, collected on different dates from 2007 through early 2009 and discussed in detail later in this report, were available for most of the study area. Where LiDAR data were not available, aerial imagery collected on June 23, 2009, and discussed in detail later in this report, was used to construct the river centerline to the upper extent of the study area at Glen Avon Bridge.

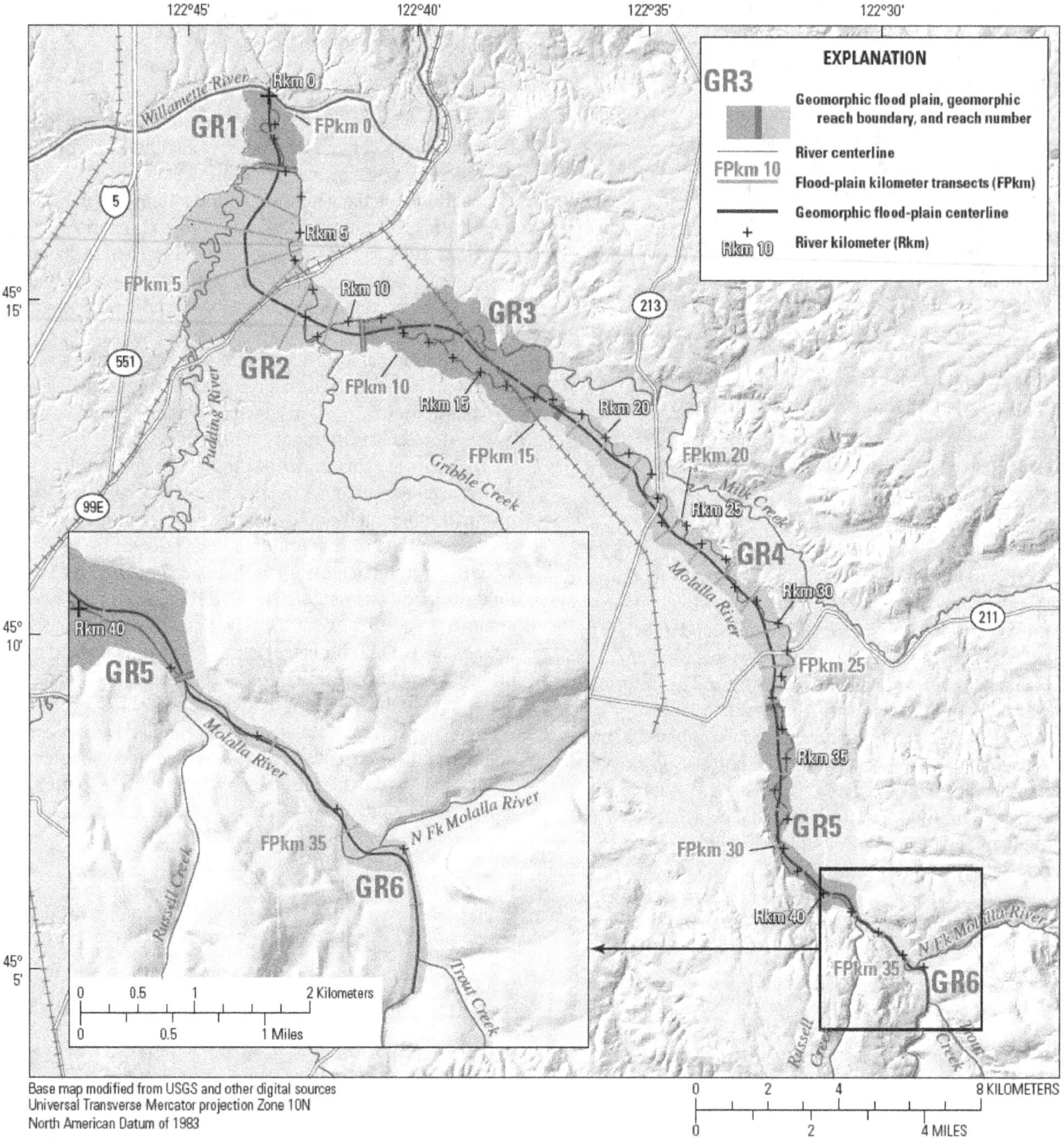

Figure 2.   Geomorphic flood plain, river centerline and river kilometer stationing (Rkm), geomorphic flood-plain reaches, and geomorphic flood-plain centerline and flood-plain kilometer transect stationing (FPkm), Molalla River, Oregon.

Table 1. Conversion between river-centerline stationing, U.S. Geological Survey (USGS) topographic map river-mile stationing, and geomorphic flood-plain stationing for the Molalla River, Oregon.

River kilometer (Rkm) centerline position	River mile (RM) location from USGS topographic maps	Geomorphic flood-plain kilometer (FPkm) location
0.0	0.0	0.0
1.0	0.4	0.8
2.0	0.9	1.2
3.0	1.4	2.1
4.0	2.0	2.5
5.0	2.6	3.6
6.0	3.3	5.7
7.0	3.9	7.1
8.0	4.5	7.3
9.0	5.1	7.8
10.0	5.8	8.5
11.0	6.4	9.6
12.0	7.0	10.2
13.0	7.7	10.9
14.0	8.6	12.1
15.0	9.2	12.9
16.0	10.0	13.7
17.0	10.8	14.6
18.0	11.3	15.1
19.0	11.9	15.9
20.0	12.6	16.8
21.0	13.2	17.7
22.0	13.8	18.5
23.0	14.5	19.1
24.0	15.1	19.7
25.0	15.5	20.3
26.0	15.9	21.0
27.0	16.4	21.7
28.0	16.9	22.4
29.0	17.5	23.1
30.0	18.0	23.9
31.0	18.6	24.7
32.0	19.2	25.3
33.0	19.8	25.9
34.0	20.4	26.8
35.0	21.0	27.6
36.0	21.7	28.5
37.0	22.3	29.3
38.0	23	30.1
39.0	23.5	30.9
40.0	24.1	31.8
41.0	24.7	32.9
42.0	25.3	33.8
43.0	26	34.8
44.0	26.7	35.6
44.2	26.8	35.7

Longitudinal location along the river corridor using a geomorphic flood-plain coordinate system is expressed as a geomorphic flood-plain kilometer (FPkm) distance upstream of the confluence of the Molalla River with the Willamette River (fig. 2; table 1). The geomorphic flood plain was delineated using methodology developed by O'Connor and others (2003). A combination of LiDAR data, soil maps (Soil Survey Staff, 2008), and Quaternary geologic maps (O'Connor and others, 2001) was examined to identify and delineate the region modified by the river during the recent (Holocene epoch, which encompasses approximately the past 10,000 years) climatic regime. A centerline drawn along the center of the geomorphic flood plain from the upper extent of the study area to the Molalla River's confluence with the Willamette River and transects drawn orthogonally to the geomorphic flood-plain centerline define the FPkm coordinate system (fig. 2). The geomorphic flood plain used herein should not be considered a regulatory flood plain, and has no relation to existing regulatory flood-plain maps or regional channel-migration mapping. The relation between the river and geomorphic flood-plain coordinate systems, and the relation between these two coordinate systems and river mile (RM) convention available as tick marks on USGS topographic maps, are shown in table 1.

Study Area and Sampling Locations

The study area included the portion of the Molalla River from about the Glen Avon Bridge, located just upstream of the confluence of the Molalla River with the North Fork Molalla River, downstream to the confluence with the Willamette River near Canby (fig. 1). Field data for the study were collected at several sites during a series of float trips down the river corridor as well as at specific sampling locations along the river. The location of the point-specific data-sampling sites for this study is shown relative to Rkm and FPkm stationing in figure 3. The specific type and timing of data collected at each sampling site are listed in table 2. Methodological details of the data collection for the study are discussed in the analysis section of the report.

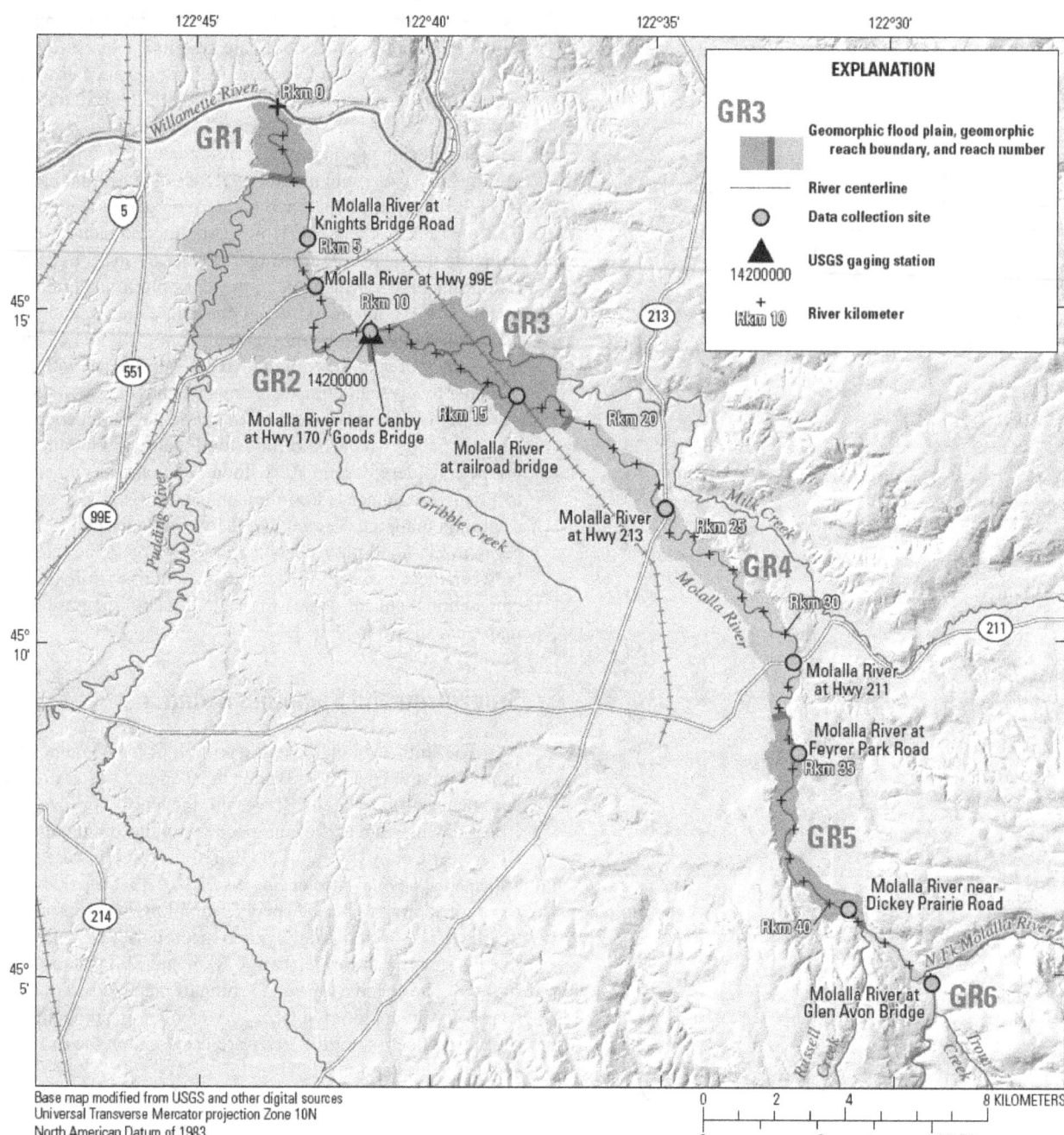

Figure 3.   Locations of point-specific data collection sites along the Molalla River, Oregon.

Table 2. Sites sampled on the Molalla River, Oregon, and details of specific data collected during July–October 2010.

[Abbreviations: Rkm, river kilometer; DD, decimal degrees; USGS, U.S. Geological Survey; m, meter; N/A, none available]

Site/reach name	USGS site identifier	Data type	Date collected	Approximate Rkm	Latitude north (DD)	Longitude west (DD)	Elevation (m)	Geomorphic reach
Molalla River at Glen Avon Bridge	450453122291000	Wolman pebble count	Sept. 2, 2010	44.32	45.0800	122.4857	163	GR6
		Water quality-nutrients and field parameters	Aug. 2–3 and Sept. 14–15, 2010	44.20	45.0811	122.4860	163	
		Benthic algae	July 30 and Aug. 15, 2010					
		Cross section survey	Sept. 2, 2010	44.19	45.0811	122.4862	162	
		Wolman Pebble count	Sept. 2, 2010	44.12	45.0818	122.4862	161	
Molalla River near Dickey Prairie Road	N/A	Wolman pebble count	Sept. 8, 2010	40.58	45.0997	122.5162	146	GR5
Molalla River at Feyrer Park Road	N/A	Cross section survey	Sept. 2, 2010	34.43	45.1393	122.5338	112	
		Wolman pebble count	Sept. 2, 2010	34.42	45.1394	122.5339	112	
Molalla River at Highway 211	450943122320400	Wolman pebble count	Sept. 2, 2010	31.11	45.1610	122.5348	97	GR4
		Water quality-nutrients and field parameters	Sept. 2, 2010	31.07	45.1612	122.5353	97	
		Benthic algae	July 30 and Aug. 15, 2010					
		Cross section survey	Sept. 2, 2010	30.92	45.1625	122.5351	96	
Molalla River at Highway 213	451200122345800	Water quality-nutrients and field parameters	Aug. 2–3 and Sept. 14–15, 2010	23.40	45.1991	122.5780	72	
		Benthic algae	July 30 and Aug. 15, 2010					
		Wolman pebble count	Sept. 2, 2010	22.94	45.1999	122.5819	70	
		Cross section survey	Sept. 1, 2010	22.84	45.2001	122.5832	70	

Table 2.   Sites sampled on the Molalla River, Oregon, and details of specific data collected during July–October 2010.—Continued

[Abbreviations: Rkm, river kilometer; DD, decimal degrees; USGS, U.S. Geological Survey; m, meter; N/A, none available]

Site/reach name	USGS site identifier	Data type	Date collected	Approximate Rkm	Latitude north (DD)	Longitude west (DD)	Elevation (m)	Geomorphic reach
Molalla River at railroad bridge	N/A	Wolman pebble count	Oct. 4, 2010	16.18	45.2280	122.6334	50	GR3
Molalla River near Canby at Highway 170 / Goods Bridge	14200000	Wolman pebble count	Sept. 8, 2010	10.74	45.2448	122.6837	34	
		Water quality–nutrients and field parameters	Aug. 2–3 and Sept. 14–15, 2010	10.50	45.2444	122.6867	33	
		Benthic algae	July 30 and Aug. 15, 2010					
		Cross section survey	Sept. 2, 2010	10.36	45.2444	122.6885	33	GR2
Molalla River at Highway 99E	N/A	Wolman pebble count	Sept. 8, 2010	6.72	45.2547	122.7063	27	
Molalla River at Knights Bridge Road	451603122423301	Water quality–nutrients and field parameters	Aug. 2–3 and Sept. 14–15, 2010	4.75	45.2688	122.7103	23	
		Benthic algae	July 30 and Aug. 15, 2010					

Description of Molalla River Basin

An understanding of the underlying geology, hydrology, and aquatic ecology of the Molalla River basin was critical to achieving the objectives of this study. Those characteristics of the basin are summarized below and are described in more detail in the watershed assessments by the Bureau of Land Management and U.S. Forest Service (1999) and Cole and others (2004).

Geology

The Molalla River flows westward from the Cascade Range into the Willamette Valley, a broad geographic region bounded by the Cascade Range to the east and the Oregon Coast Range to the west that has been topographically low since about 20 million years ago (Hampton, 1973). The Willamette Valley was filled with the Columbia River Basalt Group by about 15 million years ago and subsequently defined by continued uplift to the east and west along a north-south trending syncline (Hampton, 1972; Gannett and Caldwell, 1998). This topographic depression was subsequently filled to its current surface with as much as several hundred meters of alluvium eroded from neighboring mountain ranges and from the Columbia Plateau by late-Pleistocene Missoula Flood sediments (O'Connor and others, 2001).

The Cascade Range is subdivided into two distinct geologic regions. To the east, the Oregon High Cascades is topographically higher, with relatively gentle slopes, little dissection, and permeable bedrock that holds groundwater (Jefferson and others, 2006). To the west, the Western Cascade Range, a predominantly igneous complex assemblage consisting of andesite and basalts interspersed with pyroclastic layers, is geologically older and more deeply dissected (Gannett and Caldwell, 1998) and less permeable than the Oregon High Cascades. In contrast to its neighbors to the north and south that drain the Oregon High Cascades (Clackamas and North Santiam Rivers), the Molalla River heads almost exclusively in the Western Cascade Range.

Volcanism associated with the subduction of the Juan de Fuca plate under the North American plate built the base units of the Western Cascade Range until about 35 million years ago, when the focus of volcanic activity shifted eastward to the Oregon High Cascades (Gannett and Caldwell, 1998). The upper canyon reaches of the Molalla River study area (upstream of about FPkm 29, fig. 3) lie between upland hills of Miocene and Oligocene andesite and breccias, and the river flows through a narrow bedrock corridor of Pleistocene alluvial deposits (Gannett and Caldwell, 1998) that restricts channel movement (see photographs 1 and 2, p. 12). These Pleistocene deposits consist of interlayered

units of conglomerate and sandstone created by fluvial (cobble and gravel layering) as well as lahar (volcanic debris flow) deposition from drainage networks emanating from the Western Cascade Range. These fluvial deposits seem to correlate with periods of active Cascade Range glaciation during the Pleistocene (Clark and Bartlein, 1995; O'Connor and others, 2001). Between FPkms 20 and 29, the river flows across a flood plain of late-Pleistocene and Holocene alluvial fill bounded by Pliocene and Miocene sedimentary bedrock hills (Gannett and Caldwell, 1998). Downstream of FPkm 20, the Molalla River enters the broad alluvial plain of the Willamette Valley that cuts across high terraces of Missoula Flood sediment deposited during the late Pleistocene (O'Connor and others, 2001).

Between 15,000 and 20,000 years ago, 60 to 90 megafloods (O'Connor and Benito, 2009) burst from ice-sheet-dammed Lake Missoula in northern Montana and flowed across eastern and central Washington, and then followed the course of the lower Columbia River valley to the sea (Bretz, 1925; Baker, 1973). As these floodwaters debouched from the Columbia River Gorge, a hydraulic constriction downstream of Portland, Oregon, impounded flow to an elevation at least 150 m above sea level (O'Connor and others, 2001). At least 40 floods, primarily between 12,700 and 15,000 years ago (O'Connor and others, 2001), backed up to the upland hills bounding the southern edge of the Portland Basin and spilled into the Willamette Valley, depositing sediment over the region of the present-day Molalla River flood plain. The post-Pleistocene Molalla River incised into these Missoula Flood deposits to the base level set by the Willamette River and established the contemporary geomorphic flood plain (fig. 2). The Missoula Floods entered the Willamette Valley through two low divides: the Oregon City gap to the east and the Rock Creek gap to the west (O'Connor and others, 2001). Flood waters that poured through the Oregon City gap (fig. 1), located along the course of the Willamette River between Oregon City and Canby, created a broad, spreading alluvial fan of coarse-grained sediment, on top of which the City of Canby is built. This coarse-grained deposit bounds the northern and eastern portions of the Molalla River geomorphic flood plain between FPkms 10 and 2. Similarly, flood waters poured through the Rock Creek gap (fig. 1) to deposit a coarse-grained fan to the west of Canby (O'Connor and others, 2001). This fan bounds the western edge of the geomorphic flood plain from FPkm 1 to FPkm 4. Fine-grained deposits from the Missoula Floods extend south from Canby, filling much of the Willamette Valley with a thick mantle of sand, silt, and clay (O'Connor and others, 2001). The topographically high terrace that bounds the geomorphic flood plain from FPkm 4 to FPkm 12 is capped by this finer-grained Missoula Flood sediment.

Photographs 1 and 2.   Examples of bedrock confining channel migration in the upper Molalla River, GR6. (Photographs taken by Kurt Carpenter, July 30, 2010.)

Hydrology

The flow of the Molalla River is unregulated; however, there are numerous diversions for agriculture and other uses. The major tributary systems to the lower river are the Milk Creek basin, which enters the river from the north, and the Pudding River, which enters just upstream of the confluence with the Willamette (fig. 3). Climate in the Molalla River basin is characterized by warm and dry summers while the winters are wet and mild at lower elevations with cool temperatures.

The USGS currently operates one streamflow-gaging station upstream of Canby (station No. 14200000; Molalla River near Canby, Oregon). This gaging station has been operated discontinuously since 1928, with a total of 56 years of record (1928–59; 1963–78; and 2000-present); rainfall has been recorded at this gaging station since 2001. From 1936 to 1992, the USGS also operated a streamflow-gaging station on the Molalla River above Pine Creek near Wilhoit, Oregon (14198500). The mean monthly precipitation and discharge measured near Canby, Oregon, are shown in figure 4. Precipitation data were averaged from water year 2001 to 2010. On average, the largest monthly discharge, 69 m³/s (2,420 ft³/s), occurs in January and the smallest monthly discharge, 2.9 m³/s (102 ft³/s), occurs in August. The long-term average of minimum daily discharges is 1.7 m³/s (60 ft³/s); the lowest recorded daily flow was 0.6 m³/s (22 ft³/s) in 1959. Eighty percent of the flow of the Molalla River and 73 percent of the precipitation in the basin occurs between November and April (Bureau of Land Management and U.S. Forest Service, 1999). The short lag time between precipitation and runoff is indicative of the peak flow being rainfall dominated. The upper catchment of the Molalla River basin originates in the Western Cascade Range, a relatively steep, deeply dissected range that, compared to other streams in the Cascade Range, tends to be runoff dominated (Jefferson and others, 2006). The lack of water storage in (and later release from) either a seasonal snowpack or groundwater in the upper catchment contributes to low summer flows (Conlon and others, 2005) and warm stream temperature in summer (Jefferson and others, 2004).

The Molalla River watershed commonly receives heavy winter rain associated with atmospheric moisture that originates in the tropics (Cooper, 2005). The runoff from these storms can combine with meltwater from antecedent snow cover to create extreme high flows. Examples of such high-flow events on the Molalla River include peak flows of 1974, 1996, and the December 1964 storm, often referred to as the Christmas Flood (Hubbard, 1991; Taylor and Hatton, 1999). The 10 largest peak flow events on record have all occurred between the months of November and March.

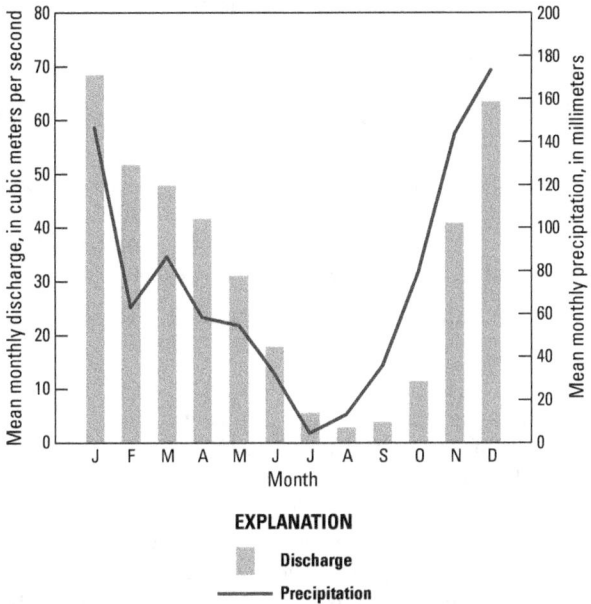

EXPLANATION

Discharge

Precipitation

Figure 4. Monthly mean precipitation and discharge of the Molalla River measured at the U.S. Geological Survey streamflow-gaging station Molalla River near Canby, Oregon (14200000).

To estimate annual peak flows for the Molalla River near Canby during years the Canby gaging station was not in operation, data from the Molalla River near Wilhoit gaging station (14198500), located at Rkm 54, and from the Clackamas River gaging station (14210000; Clackamas River at Estacada, Oregon) were used in a regression analysis of log-transformed peak-flow data. Using the annual peak-flow data from the Clackamas River gaging station ($Q_{Clackamas}$), the estimated peak flow at Canby, (Q_{Canby}), is given as:

$$\log (Q_{Canby}) = 0.76 \log (Q_{Clackamas}) + 0.80,$$

yielding a coefficient of determination of $R^2 = 0.73$. Similarly, using the annual peak-flow data from the Molalla River near Wilhoit gaging station ($Q_{Wilhoit}$), the estimated peak flow at Canby, (Q_{Canby}), is given as:

$$\log (Q_{Canby}) = 0.90 \log (Q_{Wilhoit}) + 0.68,$$

yielding a coefficient of determination of $R^2 = 0.85$. The regression estimates made on the basis of data from the Clackamas and Wilhoit gaging stations have a standard error of 80 m³/s (2,800 ft³/s) and 70 m³/s (2,400 ft³/s), respectively. Table 3 summarizes the data used to create an enhanced peak discharge record for the Molalla River near Canby for the period 1909 to 2009.

Table 3.   Summary of data used to assess the peak-discharge record for the Molalla River, Oregon.

[**Abbreviations:** Rkm, river kilometer; n/a, not applicable]

Gaging station name (or data source)	Gaging station No.	Molalla River position (Rkm)	Years of record
Molalla River near Canby	14200000	10.47	1928–59, 1963–78, 2000–10
Molalla River above Pine Creek near Wilhoit	14198500	54	1936–1992
Clackamas River at Estacada	14210000	n/a	1909–2009
Canby Fire Department	n/a	10.47	n/a

Peak flows, reconstructed in this manner, have varied between 92 m^3/s (3,250 ft^3/s) and 1,240 m^3/s (43,600 ft^3/s) between 1909 and 2009 (fig. 5). During the overlapping years of record for the two gaging stations, the estimated flows from the regression equations are plotted showing the approach reasonably approximates peak flows for those years not directly gaged. The largest peak of record on the Molalla River occurred December 22, 1964, with a measured discharge of 1,240 m^3/s (43,600 ft^3/s). This discharge produced a large flood and widespread channel changes. Other large peak flows include 1,030 m^3/s (36,200 ft^3/s) in 1972 and 883 m^3/s (31,200 ft^3/s) in 1974; both events were recorded at the USGS gaging stations. Since the Canby gaging station was reactivated in 2000, the largest peak flow was 691 m^3/s (24,400 ft^3/s), on January 2, 2009. In the past 30 years, the largest peak flow occurred on February 2, 1996, when the gaging station at Canby was not operating. Stage height at the Canby gaging station location for this 1996 high-flow event was recorded to be 7.50 m by the Canby Fire Department (Andy Bryant, National Weather Service, written commun., 2010). Although a verified stage-discharge relation to estimate discharge for the February 1996 event is not available, application of the current stage-discharge relation to this record stage would result in a discharge estimate of about 900 m^3/s (32,000 ft^3/s) for the 1996 peak event.

Using the Canby gaging record, a log-Pearson Type III (LP3) distribution was created (U.S. Geological Survey, 1981) to calculate recurrence-interval events and annual exceedance probability of peak discharge on the Molalla River (table 4).

Designated Beneficial Uses and Status of Fish Populations

Designated beneficial uses for the Molalla River include water for drinking, irrigation, and livestock; anadromous fish passage, spawning, and rearing; habitat for resident fish and aquatic life; fishing; boating; water contact recreation; and hydropower (Oregon Administrative Rules, Chapter 340, Division 41, Rule Number 340-41-0340). The Molalla River currently supports salmon and steelhead runs in varying abundances each year, although native stocks of salmonids are reduced from historical numbers, a consequence of a combination of several factors, including habitat degradation, heavy fishing pressure, reduced ocean survival, and competition with hatchery fish (Oregon Department of Fish and Wildlife, 1992; Bureau of Land Management and U.S. Forest Service, 1999). The lower reaches of the Molalla River are used by nonnative fall Chinook salmon and the upper, higher gradient reaches are used by native winter steelhead and spring Chinook (Oregon Department of Fish and Wildlife, 1992). The spring Chinook and winter steelhead are part of the Upper Willamette Evolutionary Significant Units that were federally listed as threatened under the Endangered Species Act in 1999, and reaffirmed in 2005 for spring Chinook and in 2006 for winter steelhead (National Atmospheric and Oceanic Administration Fisheries, 2005; 2006). The extinction risk for spring Chinook salmon in the Molalla River is considered high, whereas winter steelhead runs are at lesser risk (Oregon Department of Fish and Wildlife and National Atmospheric and Oceanographic Administration Fisheries, 2010). The Molalla River and some of its tributaries also provide habitat for summer steelhead, fall Chinook salmon, coho salmon, and resident cutthroat trout (McIntosh and others, 1990).

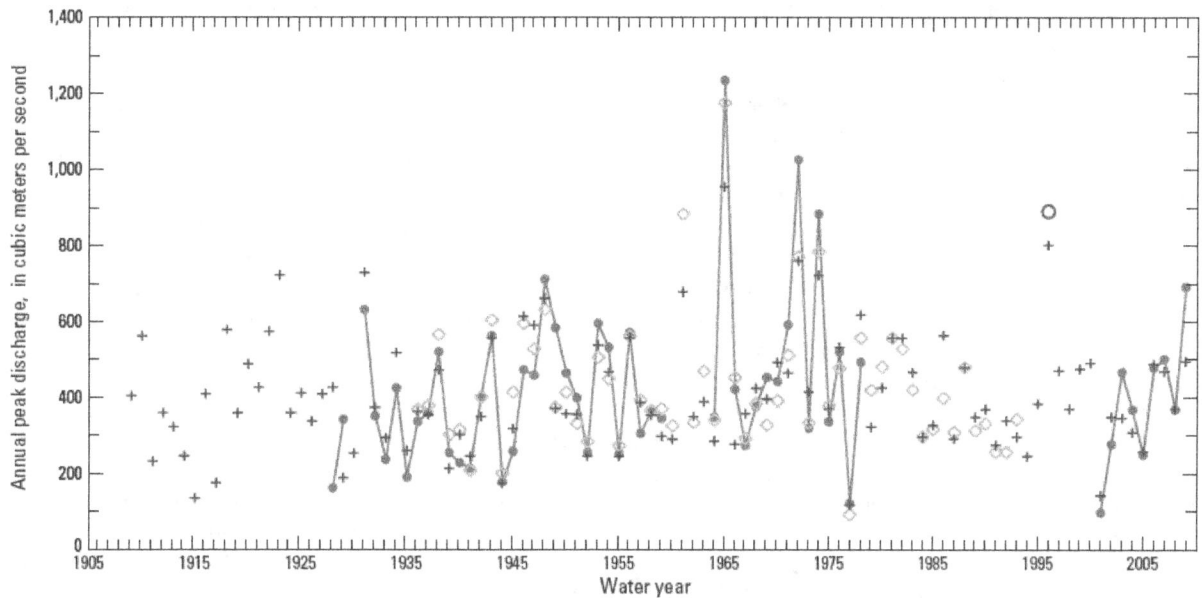

EXPLANATION

──●── Measured peak discharge; Molalla River at Canby, Oregon
        (14200000)

○   Estimated peak discharge using fire department
     observed 1996 high-water mark recorded at Canby

◇   Estimated peak discharge using regression with peak-
     flow data from Molalla River at Wilhoit (14198500)

+   Estimated peak discharge using regression with peak-
     flow data from Clackamas River (14210000)

Figure 5.   Enhanced annual peak streamflow for the Molalla River near Canby, Oregon, 1909–2009.

Table 4.   Discharge values of given annual exceedance probabilities for peak flows for the Molalla River using data from U.S. Geological Survey gaging station at Canby, Oregon.

[**Abbreviations:** m^3/s, cubic meter per second; ft^3/s, cubic foot per second]

Annual exceedance probability	Recurrence interval (years)	Discharge (m^3/s)	Discharge (ft^3/s)
0.5	2	379	13,400
0.2	5	570	20,100
0.1	10	705	24,900
0.04	25	885	31,300
0.02	50	1,020	36,200
0.01	100	1,170	41,300

Aquatic Habitat and Water Quality Conditions

The quality of aquatic habitats in the Molalla River, including riffle spawning areas, pool frequency and depth, channel complexity, and the amount of large wood is critical for resident and anadromous fish. High-quality aquatic habitats are also necessary for supporting benthic invertebrate populations that are important components of river food webs that support salmonid fishes.

Aquatic habitat quality in the Molalla River basin is variously affected by anthropogenic activities in the basin and along the river and its flood plain. Logging practices, including construction of splash dams and log drives, once scoured streambeds and fish spawning areas. Later, commercial timber harvesting and road construction probably increased sediment loading to the river that degraded habitat and increased turbidity (Bureau of Land Management and U.S. Forest Service, 1999). In 1999, the Bureau of Land Management (BLM) watershed analysis concluded that instream habitat conditions in the upper Molalla River basin upstream of Feyrer Park Road are generally fair to good in the main stem and poor to fair in most of the tributaries surveyed, based on pool frequency and size, percent secondary channels, and wood volume (Bureau of Land Management and U.S. Forest Service, 1999). The primary impacts in the upper basin include sedimentation, partly due to landslides, and lack of instream wood. Wood serves to create and maintain complex habitats, increases the retention of spawning gravel and nutrients, reduces the velocity of high flows, and creates refuge areas for juvenile and adult fish. The watershed analysis also concluded that conditions in the upper basin should improve as timber harvest units revegetate and stabilize.

In many reaches of the Molalla River, high-quality habitat exists for fish and other aquatic life, including highly productive riffles, braided channels with side channels, and relatively deep holding pools. During summer, however, some of the habitat is marginalized due to the low streamflows that contribute to high water temperatures. Some stretches of the lower river contain large rafts of wood and log jams, but much of the wood is perched at the heads of gravel bars, out of the water and inaccessible to fish. Channelization and bank stabilization projects also have the potential to impact habitat for fish by increasing water velocities or restricting access to side channels, for example, which are important refuges for fish during high flows.

Water quality is also critically important for fish that inhabit the Molalla River, and previous studies (Oregon Department of Environmental Quality, 1988; Bureau of Land Management and U.S. Forest Service, 1999; Williams and Bloom, 2008) have documented the occurrence of several water-quality issues including high water temperatures, elevated levels of bacteria and sediment, and low concentrations of dissolved oxygen associated with nutrient enrichment and growths of attached benthic algae (periphyton). Although water temperatures are problematic throughout most of the river, nutrient, algae, and dissolved oxygen conditions are most severe in the lower reaches of the river, downstream of agricultural and urban sources.

Both point and especially non-point source pollution have the potential to negatively affect water quality in the Molalla River. The wastewater facility for the City of Molalla discharges treated effluent to the river downstream of Feyrer Park during the wet season only, typically between November 1 and April 30, and uses managed land application during the dry season (Tetra Tech and KCM, Inc., 2000). Runoff from timber harvest and agricultural areas, as well as from urban areas, also contributes sediment, nutrients, and other pollutants to the river.

The Oregon Department of Environmental Quality (ODEQ) has been monitoring the quality of water in the lower Molalla River several times per year since the early 1980s as part of their ambient monitoring program, and currently collects samples at RM 3 downstream of Knights Bridge. In an early assessment of the Molalla River, ODEQ determined that water quality in the upper basin was affected by inputs of sediment from landslides, forestry, and road related runoff (Oregon Department of Environmental Quality, 1988). Water quality in the middle river, downstream of the town of Molalla, was affected by erosion and turbidity, low levels of dissolved oxygen, sediment, and low-flow problems as a result of water withdrawals (Bureau of Land Management and U.S. Forest Service, 1999). In the lower river, high water temperatures, high biochemical oxygen demand, and low levels of dissolved oxygen have been a problem during summer low-flow periods as a result of non-point source pollution (Cude, 1996). In 1998, the lower Molalla River, from its mouth to the confluence with North Fork Molalla River, was on the 303(d) list of water quality impaired streams for high water temperatures (summer) and E. coli bacteria (autumn-spring). ODEQ also noted that instream water rights were often not met at the Goods Bridge gaging station during summer, so flow modification was listed as contributing to high water temperatures.

The most recent (1998–2007) analysis of the ODEQ data classified water quality at the Knights Bridge site as "good" (score of 89 out of 100), based upon an index that includes water temperature, dissolved oxygen levels, biochemical oxygen demand, pH, and the concentrations of total solids, ammonia, nitrate, total phosphorus, and bacteria (Oregon Department of Environmental Quality, 2008). Nevertheless, several stream reaches of the Molalla River currently do not meet water-quality standards for water temperature and bacteria (Williams and Bloom, 2008). A detailed analysis of data for the temperature Total Maximum Daily Load (TMDL), including Thermal Infrared Radiometry (TIR) data collected in July 2004, found temperatures in the main stem as much as about 26°C (Williams and Bloom, 2008). Heat-source modeling (Boyd and Kasper, 2003) showed that heating resulted from increased solar warming as a consequence of a paucity of streamside vegetation, reduced streamflow, and inputs of relatively warmer water from certain tributaries.

Inputs of cooler water from other tributaries such as Table Rock Fork Molalla River, Trout Creek, North Fork Molalla River, Russell Creek, and Milk Creek produces drops in water temperatures in the main stem, but TIR data show downstream temperatures continued to rise until the next tributary downstream provided additional cooler water. Inputs of cold spring water also enter the Molalla River, mainly upstream of Gawley Creek and downstream of the North Fork Molalla River (Williams and Bloom 2008), which probably keep temperatures in the main stem from getting even warmer.

High water temperatures are stressful for salmonids and favor production of less desirable warm water fish species, some of which are introduced (Oregon Department of Fish and Wildlife, 1992). Oregon Department of Fish and Wildlife found that low summer flows and warm water temperatures were the two most important limiting factors affecting spring Chinook and winter steelhead in the Molalla River (Bureau of Land Management and U.S. Forest Service, 1999). Most (78 percent) of the water use in the Pudding-Molalla River basin is for agriculture, with 7 percent for municipal uses. Although instream flow requirements do exist in the Molalla River for aquatic life and pollution prevention, these tend to be among the most junior water rights (Williams and Bloom 2008).

As required by the Clean Water Act of 1972, and because certain stream reaches of the Molalla Pudding Subbasin were not meeting water-quality standards, ODEQ produced a TMDL and developed a Water-Quality Management Plan (Williams and Bloom, 2008) designed to protect human health, aquatic life, and other designated beneficial uses of the water. In December 2008, the U.S. Environmental Protection Agency approved the TMDL. The plan establishes waste load allocations for point and non-point sources for parameters not currently meeting water-quality standards.

Although the Molalla River is not currently on the 303(d) list for nuisance algae, high pH, or low dissolved oxygen, nutrient enrichment and these associated eutrophication issues have been a concern in the past. Excessive biomass of periphyton algae in the Molalla River and high rates of algal photosynthesis produce supersaturated levels of dissolved oxygen and high pH to levels that could be harmful for fish and other aquatic organisms. Water quality is critically important for humans as well as aquatic life because the cities of Molalla and Canby use the Molalla River for their public drinking-water supply. Excessive amounts of algae can affect water supplies by clogging intakes, producing taste and odor problems, and enriching raw water with organic carbon that can produce toxic disinfection by-products (DBPs) when the water is treated. Although DBPs were once an issue at the Canby drinking-water treatment plant, upgrades to the plant, including elimination of pre-chlorination, greatly reduced levels of DBPs in treated water. Clogging of the intake screens by algae, however, continues to be a problem during the late summer (Brian Hutchins, Canby drinking-water plant facility manager, oral commun., 2011).

Pesticides also have the potential to affect water quality in the Molalla River, although data are limited to just a few samples. In 1994, USGS identified several pesticides in the Molalla River at Knights Bridge, including the herbicides atrazine, simazine, EPTC, and napropamide, and the insecticide chlorpyrifos (USGS NWIS database, http://waterdata.usgs.gov/nwis). The lower three kilometers of the Molalla River are also influenced by the often highly turbid Pudding River, which is currently on the ODEQ's 303(d) list as water-quality impaired due to high water temperatures, fecal coliform bacteria, and elevated concentrations of legacy insecticides dieldrin and DDT, which are no longer in use (Williams and Bloom, 2008). A previous USGS study by Rinella and Janet (1998) documented high concentrations of sediment, nutrients, and several pesticides in the Pudding River and many of its agricultural tributaries. It is not known how anadromous fish returning to the Molalla River might be affected by exposure to poor water-quality conditions in the affected reach downstream of the Pudding River, or if anadromous fish homing behavior is disrupted by exposure to pesticides (Scholz and others, 2000).

Land Use and Anthropogenic Impacts

Forest management and road construction in the upper basin, and agriculture, gravel mining, urban development, and other activities in the lower basin all have the potential to affect the landscape, aquatic habitat, and water quality of the Molalla River (Bureau of Land Management and U.S. Forest Service, 1999; Cole and others, 2004). Although Federal lands owned by the BLM and USFS comprise about 35 percent of the basin, 53 percent is private industrial forestland, and the remaining land is used primarily for agriculture, interspersed by parcels of rural residential, urban, and some industrial land in the lower basin.

The Molalla River basin has a history of logging dating back to the early 1900s (see review in Cole and others, 2004). Early logging operations used the Molalla River and Milk Creek for transporting logs to mills, and included the use of log ponds, flumes, and splash dams. Some of the early practices were damaging to aquatic ecosystems as logs were skidded through streams and across fish spawning areas. In the early 1940s, 300–600 log trucks per day traveled on the Molalla Forest Road on their way to mills in Molalla (Cole and others, 2004). Widespread road construction and commercial timber harvesting began in the 1950s and 1960s, with harvest occurring down to the river's edge, which reduced wood recruitment to the river and assemblages of large woody debris in the flood plain (Swanson and Lienkaemper 1978, Likens and Bilby 1982). Today, just 4 percent of the old growth forest remains in the Molalla River basin, much of it in the Table Rock Wilderness (Bureau of Land Management and U.S. Forest Service, 1999). Intensive logging subsided in the 1970s and 1980s, and by the 1990s, most of the timber harvest had shifted to private forestland.

By 1998, there were over 1,000 km of roads in the Molalla River basin, and the dominant land cover was regeneration forests—open sapling/brush from 10 to 40 years of age and 40 to 80-year-old closed canopy stands. Water Availability Runoff (WAR) hydrologic models have shown that such removal of forest canopy and creation of extensive open areas in the upper Molalla River watershed has resulted in additional snow accumulation that has exacerbated flooding during rain-on-snow events (Bureau of Land Management and U.S. Forest Service, 1999), such as those that occurred in 1964 and 1996. Overall, the BLM's watershed analysis rated the basin to be in a poor to fair condition, and concluded that aquatic habitat in the upper Molalla River basin was now degraded from decades of forest management, timber harvesting, road building, and stream cleaning, all of which have resulted in a lack of wood in the upper river. In contrast, large accumulations of wood and large log jams are present in the middle and lower Molalla River, where they provide cover for fish and may contribute to migration of the river's channel. Property damage from historical flooding has, over time, resulted in construction of several bank stabilization projects beginning in 1938 by the U.S. Army Corps of Engineers (USACE), and more recently involving the Natural Resources Conservation Service (NRCS) and other private and public groups. Twenty of these projects have been built within the lower 40 km of the Molalla River (Cole and others, 2004).

Geomorphology

The fluvial geomorphology of the Molalla River, which demonstrates features typical of many rivers draining the Western Cascade Range, has unique characteristics and fluvial responses that warrant a focused investigation. The geomorphology of the Molalla River was analyzed using available remotely sensed data sets, including LiDAR and aerial imagery, as well as field-collected data.

Observations from reconnaissance trips to the Molalla River during the early stages of this study led to hypotheses that guided subsequent geomorphic analysis. For example, fresh gravel deposits and abandoned channels upstream from Canby (GR3, fig. 2) suggested increased channel-migration activity might have resulted from excess sediment load arriving from upstream. Elevated gravel and cobble bars upstream of the Highway 211 Bridge suggested the possibility of increased sedimentation in the upstream reaches of the study area. It was hypothesized that the river corridor might be subject to widespread aggradation, which, in turn, was leading to increased flooding and channel migration. Subsequent analysis showed systematic aggradation was not widespread and observed sedimentation was due to localized effects and proximity to channel-constraining bedrock.

Geomorphic Reaches

The geomorphic flood plain of the Molalla River was delineated to provide a static reference frame for analysis of temporal changes in the channel morphology. The geomorphic flood plain was also intended to represent the spatial extent of the lateral movement of the river during the Holocene, typically spanning the valley bottom between bedrock or alluvial bluffs that act to confine channel movement. In analyzing both the geomorphic flood plain and other geomorphic and physical characteristics of the river corridor, six reaches exhibiting unique physical traits were identified. These unique geomorphic reaches were determined using LiDAR-derived hill-shade maps that revealed distinct transitions in channel width, presence of secondary channels, and channel-migration potential.

For the Molalla River, two separate bare-earth LiDAR data sets were obtained through the Oregon Department of Geology and Mineral Industries. LiDAR data spanning Rkm 0.0–22.36 (northern region) were collected between March 16, 2007, and April 15, 2007; LiDAR data encompassing Rkm 22.52–39.41 (southern region) were collected on October 26, 2008, November 1-4, 2008, and March 18, 2009 (Watershed Sciences, 2009a; 2009b). The LiDAR data sets contained topographic points at 0.91-m grid resolution, with vertical accuracy of 5-8 cm in the northern region and 1-3 cm in the southern region.

Interfaces between geomorphic reaches were also determined by identifying narrow points in the geomorphic flood plain (fig. 6) that indicated natural transitions. The geomorphic flood plain within an individual reach tended to have a relatively consistent width.

The geomorphic reaches were numbered from one to six from the confluence with the Willamette River to the upstream extent of the study area (table 5; fig. 3). Geomorphic reach 1 (GR1) extended from the Willamette River (FPkm 0.0) to the confluence of the Molalla River with the Pudding River (FPkm 2.0). Geomorphic reach 2 (GR2) extended from the confluence with the Pudding River to Goods Bridge at Highway 170 (FPkm 9.0) just south of Canby and adjacent gaging station, Molalla River near Canby (14200000). Geomorphic reach 3 (GR3) extended from Goods Bridge to a relatively narrow point in the geomorphic flood plain at FPkm 15.4 (fig. 6) located just upstream of where the Milk Creek tributary enters the Molalla River geomorphic flood plain. Geomorphic reach 4 (GR4) extended from FPkm 15.4 upstream to another relatively narrow point in the geomorphic flood plain (FPkm 26.1). Geomorphic reach 5 (GR5) extended from FPkm 26.1 to FPkm 33.0, where the overall geomorphic flood plain narrows markedly (fig. 3 and fig. 6), and geomorphic reach 6 (GR6) encompassed the remainder of the study area upstream to Glen Avon Bridge at FPkm 35.7.

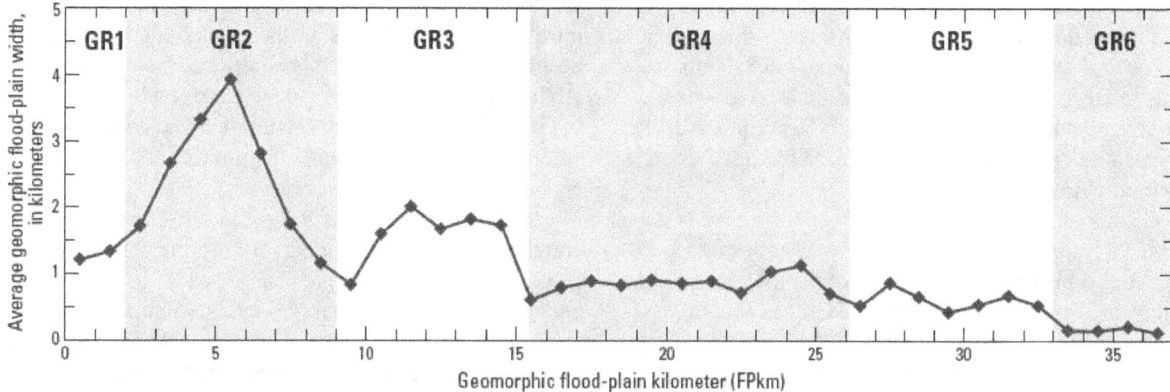

Figure 6.   Geomorphic flood-plain width along the Molalla River, Oregon.

Table 5.   Upstream and downstream extents of the six geomorphic reaches along the lower Molalla River, Oregon.

[**Abbreviations:** FPkm, flood plain kilometer; Rkm, river kilometer]

Reach name	Geomorphic flood plain		River centerline	
	Downstream extent (FPkm)	Upstream extent (FPkm)	Downstream extent (Rkm)	Upstream extent (Rkm)
GR1	0	2	0	2.89
GR2	2	9	2.89	10.39
GR3	9	15.4	10.39	18.43
GR4	15.4	26.1	18.43	33.25
GR5	26.1	33.0	33.25	41.13
GR6	33.0	35.7	41.13	44.19

Channel Characterization within Geomorphic Reaches

The upstream-most geomorphic reach (GR6) is a relatively narrow, 2.7-km-long fluvial corridor constrained by sandstone and conglomerate bedrock on both banks. Numerous bedrock exposures in the channel bottom throughout GR6 indicate that alluvial fill is thin and the river has incised into bedrock in recent geologic history or is currently downcutting. As a result of bedrock control, there are no secondary or side channels in GR6; however, pools are frequent in GR6, commonly eroded directly into exposed bedrock just downstream of steep riffles. Some tributaries have deposited large boulder assemblages into the main stem that appear to be eroded debris-flow or landslide deposits. From Glen Avon Bridge to the North Fork of the Molalla River (Rkm 43.8), the river has a plane-bed morphology, that is, a relatively flat cobble-bed channel lacking discrete bars that has a low width to depth ratio and large relative roughness (Montgomery and Buffington, 1997). Downstream of the North Fork Molalla River, large floods with ample stream power have mobilized particles and reworked the

coarse-grained sediment into a pool-riffle morphologic structure (Montgomery and Buffington, 1997) with boulder bars establishing and controlling the riffle structure. Boulders are the dominant sediment-size class of the large channel-forming bars. LiDAR data were not available for GR6, which precluded a complete analysis of slope, although a coarse calculation from USGS topographic maps showed reach slope to be about 0.005 m/m. A relatively small number of engineered revetments in GR6 mantle bedrock to protect structures built on top of adjacent strath terraces but do not restrict channel movement any more than the underlying bedrock. The narrow, constrained river corridor in GR6, combined with established and mature riparian vegetation, results in a well-shaded channel with riffles and pools (fig. 7A) that keep the water well oxygenated. Despite the established riparian corridor, large woody debris in GR6 is rare due to limited recruitment from channel migration and efficient transport from the reach during high flows.

Leaving the constrained bedrock corridor of GR6, the river enters the alluvial, 6.9-km-long GR5. The mean river slope is about 0.005 m/m. Compared to GR6, the active channel of GR5 is wider and riparian vegetation is farther from the river centerline, resulting in less overall shade. Compared to downstream reaches, however, GR5 is relatively narrow and well shaded (fig. 7B). The morphologic structure of GR5 is pool-riffle, but sediment particle size in the channel-forming bars declines progressively in the downstream direction: fluvially deposited boulder bars are numerous in the upper extent of GR5 and rare in the lower extent; gravel and cobble deposits are common throughout the reach. Sandstone and conglomerate bedrock crops out along the channel bed throughout GR5, but less abundantly than in GR6. Similarly, laterally constraining bedrock occurs sparsely and only where the main stem approaches the outer extents of the wider geomorphic flood plain on both river-right and left. In contrast to GR6, a high-flow alluvial flood terrace, 1 to 4 m higher than the river, is prevalent on both sides of the river throughout GR5, indicative of a river with some degree of active channel migration over the recent geologic time frame. Revetments are infrequent in GR5, but they do restrict channel migration into the alluvial flood plain in strategic locations. Secondary channels occur in the downstream section of GR5. Pools are numerous in GR5, and the deepest pools are located where the channel approaches lateral bedrock at the periphery of the geomorphic flood plain. There is relatively little large woody debris within GR5.

In GR4, a wide and open active channel within a wider geomorphic flood plain, the Molalla River assumes the characteristics typical throughout most of the lower extent of the study area (fig. 7C). The length of GR4 is 10.7 km and the slope of the river here is about 0.0035 m/m. The same units of sandstone and conglomerate bedrock present in GR5 and GR6 underlie much of the channel throughout this reach. Lateral bedrock constrains the river in GR4 only on the right (northeast) extent of the geomorphic flood plain as the left edge of the geomorphic flood plain in GR4 is exclusively a terrace of Pleistocene alluvium. Revetments to protect property, agriculture, and infrastructure are common throughout GR4; they are prevalent on the left side of the river because confinement on the right side of the river is commonly governed by bedrock. The morphologic structure is pool-riffle throughout GR4, and pools, although present throughout the reach, tend to be deeper near confining bedrock or revetments. Particle size in alluvial channel-forming bars progressively finer downstream, with boulder bars giving way to cobble and gravel bars. Secondary channels are not widespread, but are more common than farther upstream. Shade in GR4 becomes sparse as much of the riparian vegetation along the river channel has been removed, exposing more of the water surface to direct sunlight. Limited assemblages of large wood begin to appear in the lower section of GR4, sourced, presumably, from active channel migration within the reach.

As it enters the wide, alluvial geomorphic flood plain of GR3, the Molalla River is active with historical tendencies toward flooding and rapid channel migration, which presents major challenges for people living and working along its banks. The total length of this reach of the geomorphic flood plain is 6.4 km, and the mean slope along the river is about 0.003 m/m. Whereas bedrock is present in the channel of the upper 1 km of GR3 and at one location along the right bank near FPkm 14.8, bedrock is absent from the lower 5 km of GR3, indicating the river here flows over alluvium of unknown depth. The coarse-grained fan of Lake Missoula Flood deposits (O'Connor and others, 2001) in the lower 1 km of GR3, however, does restrict lateral movement of the river to the north. The morphologic structure of the river in GR3 is pool-riffle with island-braided characteristics (Beechie and others, 2006), and numerous cobble and gravel bars occupy a wide and exposed active channel (fig. 7D). Pools are frequent in GR3. Secondary channels are common, due in part to active channel movement during historical peak flows and a relative paucity of revetments. Few revetments exist in the reach and

are predominantly placed on river left to protect property and infrastructure. The largest of these is a 0.6-km-long revetment near FPkm 14 that protects the railroad line on river left. Of all the reaches, riparian vegetation and shade is least common in GR3 (fig. 7D) although recruitment and collection of large woody debris on the channel margin from channel migration appears to be an active process.

As the Molalla River flows past the city of Canby, it enters the relatively confined 7.0-km-long corridor of GR2 (fig. 7E). The overall active channel width in GR2 is narrow and the mean river slope is about 0.0016 m/m. Much of the eastern boundary of the geomorphic flood plain in the downstream section of GR2 is confined by the coarse-grained fan from Lake Missoula Floods spilling through the Oregon City gap (O'Connor and others, 2001). Away from the Lake Missoula Floods deposits, the river flows between Holocene alluvial terraces. Although geologically speaking the river probably swept across the wider geomorphic flood plain in GR2, strategically placed revetments along the river in this reach have restricted the location of the river during the past four decades. No lateral bedrock exists in GR2, but channel bedrock is common and widespread from FPkm 8 downstream to FPkm 2.6. This channel bedrock consists of weakly cemented, fluvially deposited gravels probably emplaced by a Pleistocene-era Willamette River and is distinctly different from the sandstone and conglomerate units present in GR4 and farther upstream. Although not confirmed, the exposed bedrock in GR2 probably is the Linn Formation (O'Connor and others, 2001). The morphologic structure of GR2 alternates between pool-riffle and a straight channel (Beechie and others, 2006), and gravel bars are present along the channel margins although they are not as large or as common as the gravel bars found in GR3. Pools are frequent in GR2, and the deepest pool recorded in the study (6.6 m) was located near the downstream end of GR2. Riparian vegetation is common, and the narrow channel generates more shade than in GR3. There is almost no large woody debris in GR2, possibly due to the relative lack of channel movement, which in turn reduces local recruitment.

From the confluence of the Pudding River to the Willamette River, the Molalla River is a 2.0-km-long, wide river with an underlying pool-riffle morphology (fig. 7F). Unfettered by lateral bedrock or revetments, the river meanders widely and actively across the geomorphic flood plain with large gravel point and lateral bars throughout GR1. Bedrock is not exposed in GR1. No secondary active channels were apparent in the 2009 river configuration although secondary channels were probably common throughout the Holocene. Because GR1 is relatively wide, there is little shade over the water surface, but recruitment and deposition of ample large woody debris in the channel increase ecologic complexity and provide fish habitat.

Figure 7. Typical channel characteristics for each geomorphic reach in the lower Molalla River, Oregon.

Flood-Plain Morphology

Following the methodology of Jones (2006), height above water surface (HAWS) maps were generated where LiDAR data were available to visualize fluvial features along the geomorphic flood plain. Along transects drawn orthogonally to the river centerline, elevation of the flood plain is subtracted from the presumed water-surface elevation—determined from LiDAR where the river centerline intersects the transect of interest—to create an elevation of the flood plain relative to the river's water-surface elevation. Despite inaccuracies in the underlying LiDAR data and water-surface elevation, the resulting HAWS maps, when viewed as a raster data set, allowed qualitative insight of the position of the river in the recent geologic past. Specifically, the HAWS maps enabled the identification of relict morphologic features in the flood plain created by the river, such as abandoned channels, meander bends, and oxbows.

The HAWS maps show the Molalla River has actively meandered across the width of the geomorphic flood plain in GR1, combining with the Pudding River in the lower 2 km of the river corridor (fig. 8). HAWS maps in GR2 show the distinct morphology of the Molalla River and the Pudding River. The level of sinuosity of the Pudding River, on the western side of GR2, over the late Holocene has been greater than the main-stem Molalla River. Although also meandering, the Molalla River in GR2 has not been as sinuous as its western tributary. Based on the absence of fluvial features in the flood-plain surface between the two rivers

toward the southern extent of the geomorphic flood plain, there is no evidence the two rivers have joined upstream of about FPkm 6.1 within the recent geologic past. The HAWS maps of GR3 show the dominant geomorphic influence of the railroad, and its long, fortified structure dissecting the flood plain. The HAWS maps show active movement of the river before construction of the railroad line within the upstream section of GR3. This railroad line was constructed in 1913 (Cole and others, 2004). The maps also show a pre-settlement river that swept across the flood plain with braided or island-braided characteristics. It is apparent from the HAWS maps that the Molalla River in GR3 has been active, with migrating meanders and channel avulsions (that is, the rapid abandonment of a river channel and formation of a new channel), even before European settlement. Similarly, the Molalla River in GR4 has migrated actively across the geomorphic flood plain, but the process of movement has been dominated by avulsions in a pool-riffle or island-braided morphology and less so by lateral meandering of the main channel (fig. 8). The HAWS maps in GR5 indicate that the river in this reach has been a predominantly pool-riffle or island-braided system prior to development, not unlike the river today, although contemporary channel movement in GR5 was likely suppressed by bank stabilization projects. LiDAR data were not available for GR6, precluding the generation of HAWS maps; however, field observation confirmed that the river in GR6 is bedrock controlled and not subject to active channel migration.

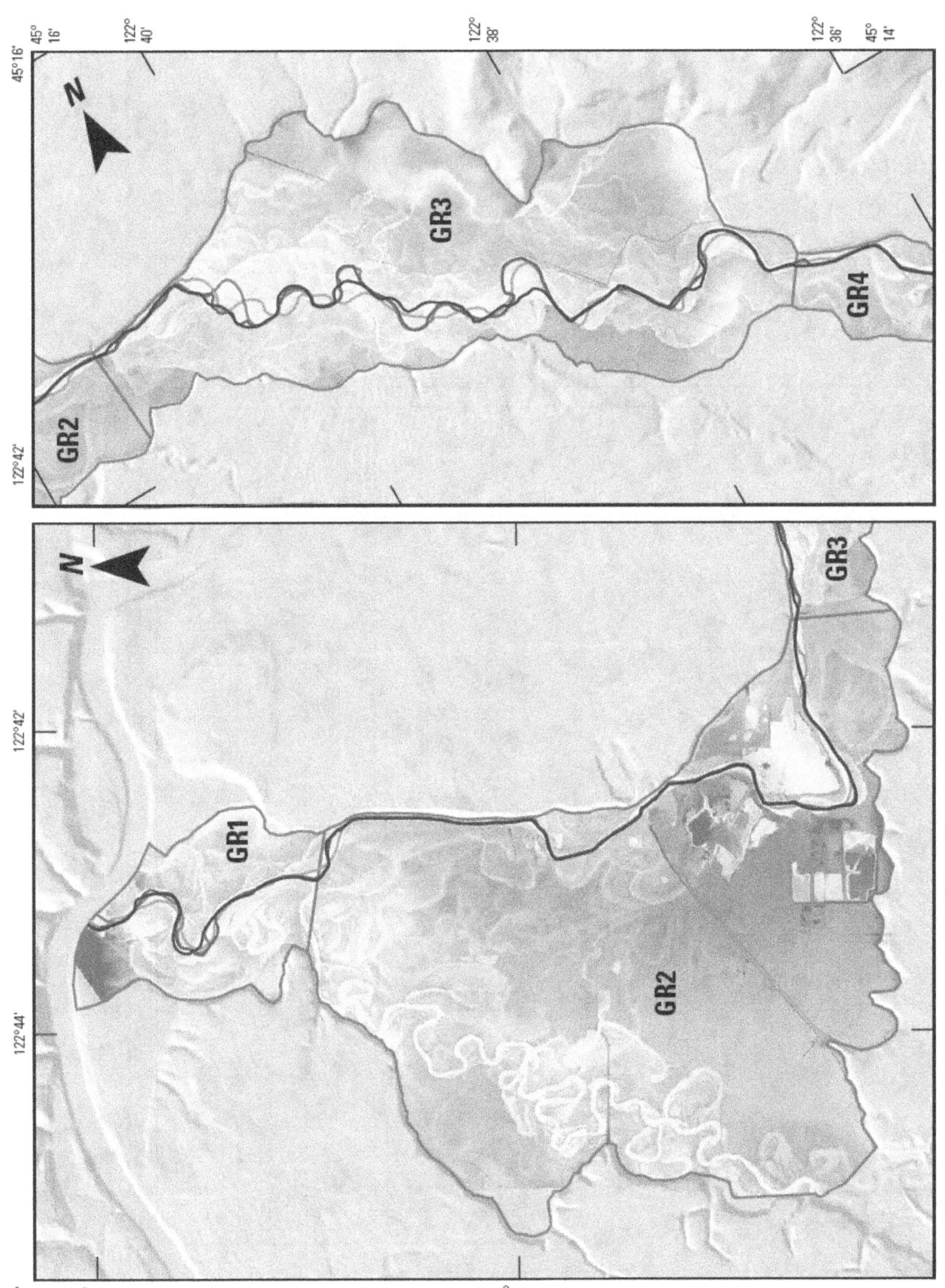

Figure 8. Height above water surface (HAWS) of the Molalla River centered on each geomorphic reach. (Using a technique to show elevation difference of the flood-plain topography relative to the river water surface [Jones, 2006], relict channels and pre-development morphologic structures of river reaches are readily identified.)

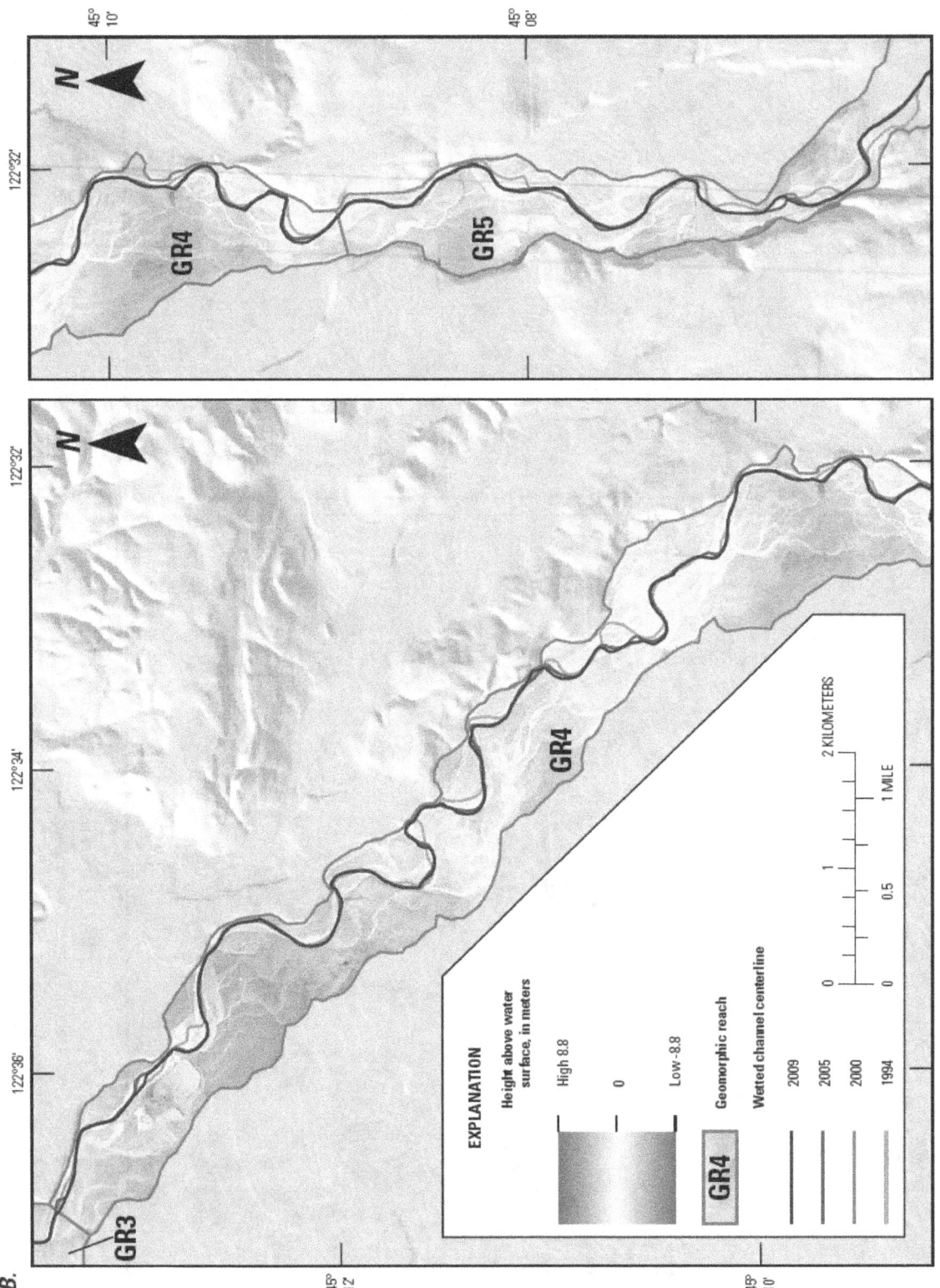

Figure 8. Continued

Key Longitudinal Trends along the River Corridor

Longitudinal trends along the river corridor were analyzed on the basis of a combination of LiDAR data, aerial imagery, and field data collection. A longitudinal profile of water-surface elevation was generated by projecting the river centerline onto LiDAR data. Three kayak float trips covering the Molalla River from Glen Avon Bridge to the Willamette River were made in July 2010 to gather river-depth data, catalogue the locations of confining bedrock and revetments, and to photograph river characteristics. Measured depths, combined with the LiDAR-derived water-surface elevation profile, were used to generate a bed-elevation profile and characterize pools in the river. Finally, field work in September 2010 provided particle-size data and bathymetric data for multiple locations in the study reach.

Water-Surface Elevation

A water-surface elevation profile was constructed from LiDAR data from Rkm 0.0 to Rkm 39.41 (fig. 9). An available 10-m digital elevation model (DEM) upstream of Rkm 39.41 was too coarse to accurately determine water-surface elevation. Airborne LiDAR systems work by firing a laser at the ground and measuring the return time of the beam reflected off the target surface (Wehr and Lohr, 1999). Although some systems were designed specifically for the purpose of mapping aqueous and subaqueous surfaces, the LiDAR flown over the Molalla River was tuned to measure terrestrial relief, and LiDAR returns from aqueous surfaces were probably discarded and topography at water features were likely interpolated from the nearest known ground points. Thus, the vertical accuracy of points in the main stem of the river is subject to errors associated with these interpolated surfaces. Differences in river discharge during different flight days over different sections of the river also create uncertainty in the final water surface. To generate a water-surface elevation profile of the river centerline, the river centerline was projected on the LiDAR data and elevation points were sampled at 1-m increments and then plotted in an x-y coordinate system. From the raw centerline elevation data, outlier points, those points that obviously misrepresented the water-surface profile, were removed, and depressions and convexities were smoothed to create a realistic profile. The vertical accuracy of the constructed water-surface profile could not be determined independently, however, the profile at any given point along the river centerline probably is within 1 m of the actual water surface. Moreover, for the analyses completed for this study, the degree of accuracy in vertical elevation is adequate to discern general trends of the water-surface profile with respect to the flood-plain scale in terms of slope, profile convexities, profile concavities, and residual pool depth.

The longitudinal water-surface elevation profile of the Molalla River (fig. 9) is generally concave, consistent with graded rivers that have adjusted their slope to sediment transport supplied from upstream (Macklin, 1948; Knighton, 1998). If the profiles are examined closely, however, they show that the Molalla River deviates from the concave profile of a graded river near Rkm 18 along the interface of GR3 and GR4 (fig. 9A). This change in curvature is more visible when the longitudinal profile is plotted on a semi-logarithmic scale (fig. 9B). The slope of the average water-surface (fig. 9C), computed as a 5-km moving average, shows the steepest portions of the river in the study area to be in the upstream extent with the gradient generally decreasing in the downstream direction to the Pudding River tributary. The increased slope at the upstream end of GR3 is also apparent in the slope data shown in figure 9C.

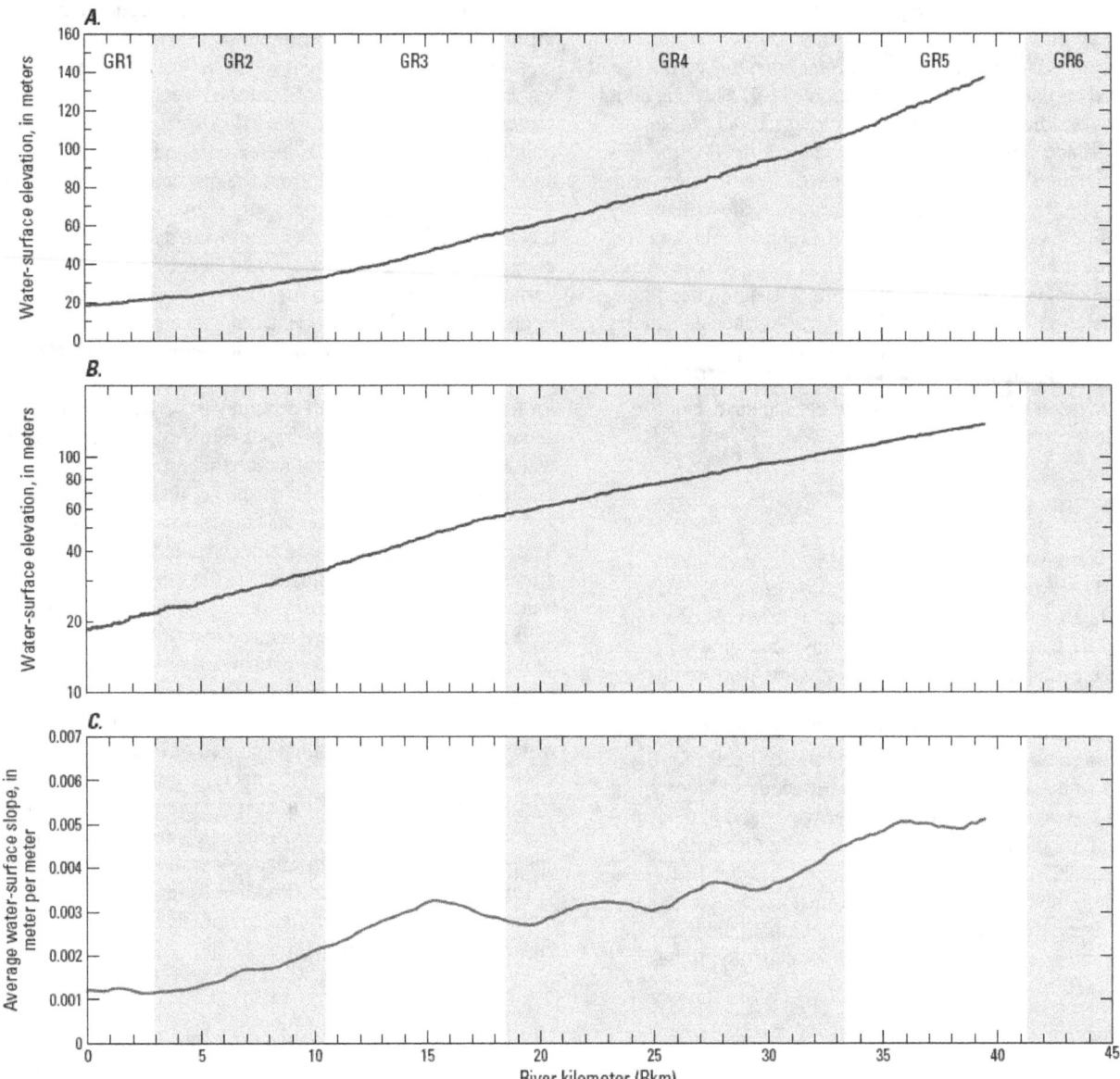

Figure 9.   Light Detection and Ranging (LiDAR)-derived longitudinal water-surface elevation profile of the Molalla River, Oregon, with elevation plotted in (*A*) normal coordinates and (*B*) semi-log coordinates. (*C*) The slope of the water-surface elevation is also shown as a function of the river centerline.

Water Depth and Residual Depths

A series of three float trips, using a two-person inflatable kayak, was completed in July 2010 to measure water depth and to map bedrock and revetments on the Molalla River from Glen Avon Bridge to the confluence with the Willamette River. The section of the Molalla River from Highway 211 to Goods Bridge was floated on July 13, 2010 (the mean discharge for the day was 6.60 m³/s, or 233 ft³/s). The section of river from Goods Bridge downstream to its confluence with the Willamette River was floated on July 14, 2010 (the mean daily discharge was 6.06 m³/s, or 214 ft³/s). The section of river from Glen Avon Bridge to the Molalla River at Highway 211 was floated on July 22, 2010 (the mean daily discharge was 4.79 m³/s, or 169 ft³/s).

During each float, water depth was measured using a global-positioning system (GPS) referenced fathometer mounted on the boat, with a specified horizontal accuracy of less than 15 m and an estimated depth accuracy of 0.2 m. The GPS-referenced water depths were recorded every 5 seconds and were limited by the instrument capability and boat dimensions to values greater than about 0.3 m. During each float, a conscious effort was made to pilot the boat as close to the thalweg (channel invert, or minimum elevation in the river bed) as possible.

During the floats, bedrock outcrops and engineered revetments were mapped using a hand-held GPS, with a horizontal accuracy of about 10 m. All observed outcrops of bedrock along the banks of the river and along the bed of the

river were identified during the float trip and categorized as being within the channel or on the left or right bank. Also, lengths of observed hardened embankments, revetments, and other man-made structures visible from the river were mapped to quantify the current degree of the anthropogenic influence on the river. It is important to note that revetments away from the river or those obscured by vegetation were not mapped; thus the revetment data presented in this report does not represent a rigorous inventory of all revetments in the river corridor.

Water-depth data measured during the study were related to the river centerline and plotted for the entire study area in figure 10. By joining the measured depth data to the nearest 1-m stationing point along the river centerline and the corresponding longitudinal water-surface elevation, a longitudinal bed-elevation profile was generated. Residual pool depth is defined as the difference between the maximum water depth of a pool and the limiting elevation of the riffle at the downstream extent of the pool (Lisle, 1987). After generating the bed-elevation profile, residual pool depth was computed from Rkm 0 to Rkm 39.41 using an algorithm and software developed by Madej (1999). The Madej algorithm identifies only pools deeper than a user-specified threshold, which was set to 1.0 m for the current study consistent with the threshold used by Madej and Ozaki (2009) for Redwood Creek, California. Redwood Creek, in the California Coastal Range where hydrology is similar to that in northwestern Oregon, drains 720 km², a drainage comparable to the Molalla River.

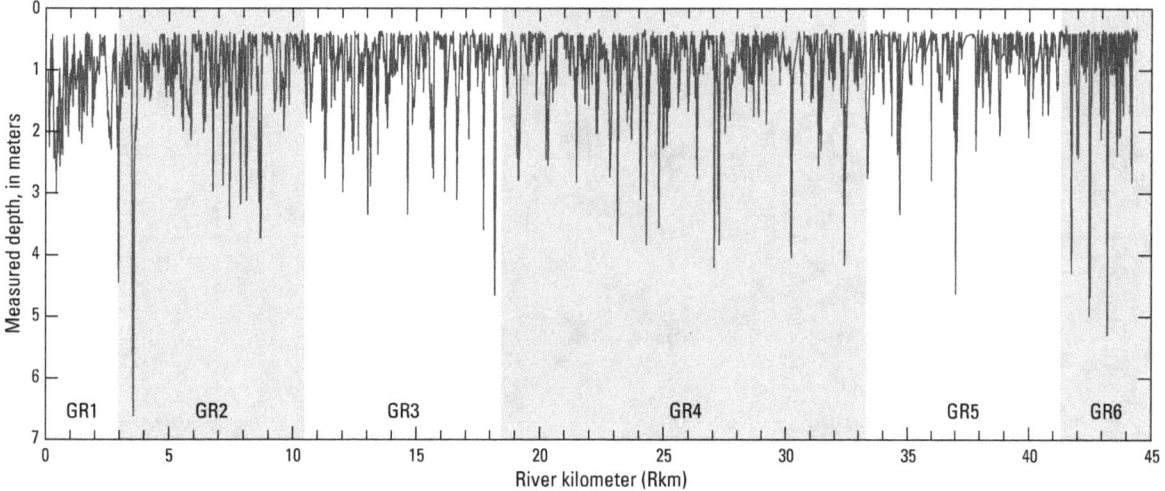

Figure 10. Water depth data from the Molalla River, Oregon, as measured in July 2010 by a fathometer mounted to boat floating down the river thalweg.

Residual pool depth statistics generated by the Madej (1999) software showed the average residual pool depth was consistent throughout the river corridor, and ranged from 1.8 to 2.1 m (table 6). Pool frequency per river kilometer ranged from 1.6 to 3.4, with the largest pool frequencies in GR1 and the smallest pool frequency in GR5. Pool frequency generally increased in the downstream direction. The maximum residual pool depth ranged from 3.6 to 6.3 m, with the maximum residual pool depth in GR2.

Table 6. Residual pool depth summary statistics by geomorphic reach in the Molalla River, Oregon, for residual pools greater than 1.0 meter.

[**Abbreviations:** m, meter; n/a, not applicable]

Geomorphic reach	Average residual pool depth (m)	Maximum residual pool depth (m)	Total pool count per geomorphic reach	Average pool frequency per river kilometer
GR1	1.8	4.0	10	3.4
GR2	2.0	6.3	17	2.3
GR3	2.0	4.3	21	2.6
GR4	2.0	3.6	29	2.0
GR5	2.1	4.2	10	1.6
GR6	n/a	n/a	n/a	n/a

River-Profile Convexities

Convexities in long river profiles are sometimes used to identify regions of alluvial accumulation (for example, Hanks and Webb, 2006) or to identify incision knickpoints (for example, Stock and Montgomery, 1999). To gain insight into morphologic evolution of the Molalla River, a second-order polynomial trend line representing a graded profile (Mackin, 1948) was fit to the longitudinal profile of water-surface elevation, and the elevation differences between the long profile and the idealized fit were plotted (fig. 11). The bed elevation profile and observations of channel bedrock are also shown in figure 11. Where observed, the channel bedrock was assumed to have the same elevation as the channel bottom, determined by water-depth measurements. The placement of bedrock points in figure 11 is not intended to represent the precise elevation of individual outcrops of bedrock; rather, the points are intended to show generally where bedrock is exposed along the river corridor. The plot shows a strong convexity centered near Rkm 18 that spans GR3 and GR4. A first interpretation might attribute this convexity to the presence of an alluvial wedge; however, the proximity of channel bedrock to the water surface from Rkm 18 to Rkm 38 shows there is little accumulated sediment upstream of Rkm 18, and that the river from GR6 downstream through GR4 flows over a terrace of bedrock positioned just below the water surface. Farther downstream, the outcrops of

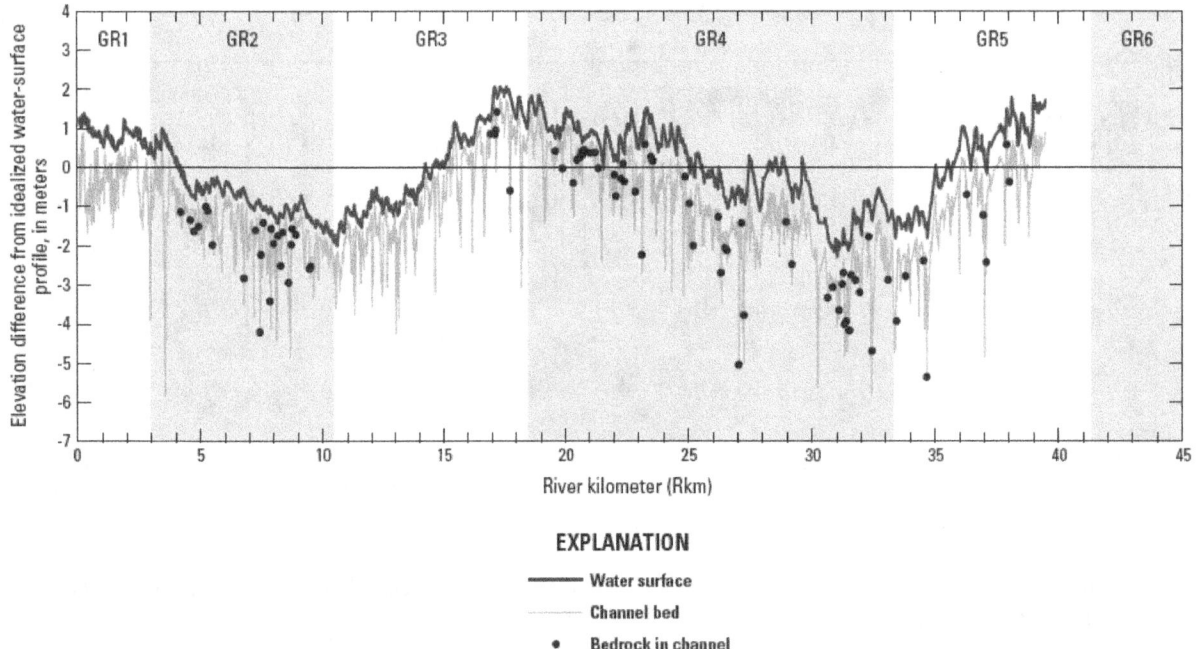

EXPLANATION

——— Water surface

········ Channel bed

• Bedrock in channel

Figure 11. Deviation of water-surface elevation profile from second-order polynomial trendline, and measured channel-bed and bedrock within the channel, Molalla River, Oregon.

channel bedrock in GR2 indicate that from Rkm 4 to Rkm 10, the river also flows over a bedrock terrace with a relatively thin mantle of alluvium. Although not determined, it appears that 2–3 m of sediment mantles bedrock in GR1, between the Pudding River and the Willamette River. Between Rkm 10 and Rkm 18, some unknown thickness of alluvium covers the bedrock, but the differenced elevation profile (fig. 11) shows no convexity between Rkm 10 and Rkm 18 that might indicate accumulating sediment. Instead, the Molalla River in this section is well graded to the long profile between bedrock outcrops, suggesting the river can effectively transport sediment from upstream through this reach. Furthermore, as a transport reach, the section of river between Rkm 10 and Rkm 18 is not accumulating new sediment to significant depths.

Though beyond the scope of this study, a possible explanation for the lack of bedrock exposures in GR3 may be the presence of a previously unmapped thrust fault near Rkm 18, where the eastern side has risen relative to the western side. Prior work by the USGS identified an area of geomagnetic anomaly near Rkm 18, possibly indicating a geologic fault (Richard Blakely, U.S. Geological Survey, written commun., 2010). Another explanation for the lack of bedrock exposures is that unique bedrock units upstream and downstream of GR3 led to differential incision rates east and west of Canby that is manifested in the river profile as a change in slope at Rkm 18. Observations during float trips that the sandstone and conglomerate units upstream of GR3 are distinctly different from the conglomerate lithology of bedrock in GR2 support this second hypothesis. The two hypotheses are not mutually exclusive and both mechanisms could have affected the underlying geologic structure.

Particle Size

Particle-size data were collected from exposed bars in September and October 2010. Using the methodology of Wolman (1954), pebble counts were conducted at nine sample locations in all geomorphic reaches except GR1 (fig. 3; table 2). At each site, prominent, unvegetated, recently deposited gravel or cobble bars near the main-stem river were selected for characterization using an approach similar to the one employed by Wallick and others (2010a). For each site, particle-size data were collected along two parallel transects placed on the sampled bars, and a total of 200 particles were measured. The pebble-count transects were typically 2-4 m from the water's edge and parallel to the flow direction of the river. The particle data were then bracketed into respective *phi* classes and analyzed (Garcia, 2008). It is important to note that sampled bars were generally at a lower elevation, thus representing bars deposited or reworked by flows within the past one to two years. As a result, the magnitudes of the flows

depositing or reworking the sampled bars are much smaller than the peak flows that control the geomorphology of the river and would deposit the highest elevation bars and coarsest particles. Thus the particle-size data presented herein well represents the clasts typically mobilized in a bankfull event.

Figure 12 shows the median particle-size (D_{50}), the 90th percentile particle-size (D_{90}), and the 10th percentile particle-size (D_{10}) data for the nine sampled bars in the Molalla River (appendix A). GR6 had the coarsest collection of particles, and median size decreased generally in the downstream direction, with GR2 having the finest particles. The farthest upstream location, just upstream of Glen Avon Bridge at Rkm 44.32, had a median particle size of 139 mm and a D_{90} of 405 mm. The bar just downstream of Glen Avon Bridge, at Rkm 44.12, had a median particle size of 76 mm and a D_{90} of 227 mm. This section of river, upstream of the North Fork Molalla River, had a plane-bed morphology with few recently deposited cobble bars. The sampled region upstream of Glen Avon Bridge, in particular, was a higher, exposed part of the main channel-forming bed composed of coarse particles and was not an accurate representation of the particles mobilized in a bankfull event. Instead, the sampled bar downstream of Glen Avon Bridge, with a smaller overall particle-size distribution, better represented the material mobilized in a bankfull event. Indeed, from this bar downstream of Glen Avon Bridge through GR3, the median grain size on the bars remained relatively constant, suggesting efficient transport, although there was a general downward trend in D_{90} (fig. 12). The one sampling location in GR2 showed a median size of 42 mm and a D_{90} of 77 mm.

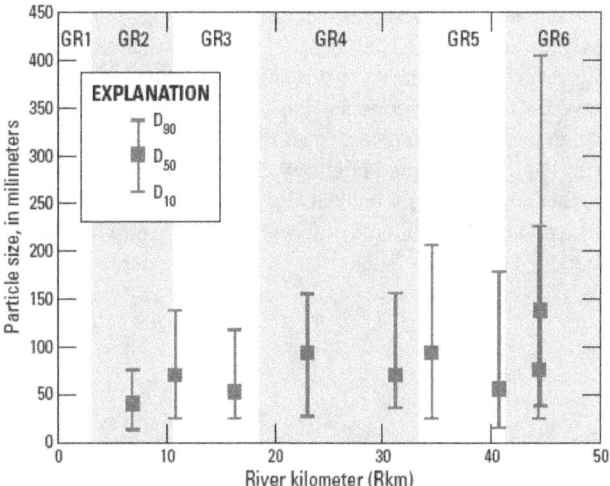

Figure 12.   Distribution of particle-size data collected using Wolman (1954) pebble counts on nine freshly deposited bars along the river corridor of the Molalla River, Oregon.

Sand-size particles (diameter less than 2 mm) were noted in the pebble counts, although not counted in the computations of particle-size statistics. Sand counts in the eight upstream-most pebble-count locations were less than 2 percent. The farthest downstream sample site in GR2 had a sand count of 5.5 percent. Overall, the Molalla River is a unimodal, gravel-bedded system with little sand.

Cross-Section Data

During September 2010, five bathymetric cross sections were surveyed near bridges throughout the study area to characterize channel geometry of the Molalla River. The cross section near Highway 213 was surveyed on September 1, 2010 (the mean daily discharge was 2.49 m³/s, or 87.9 ft³/s). The other cross sections (at the Glen Avon Bridge, Feyrer Park Road, Highway 211 Bridge, and Goods Bridge) were surveyed on September 2, 2010, when the mean daily discharge was 3.40 m³/s, or 120 ft³/s. All cross sections were surveyed using a total-station survey instrument and reflective prism on a stadia rod. The relative accuracy between surveyed points was about 0.03 m. The five cross sections were surveyed relative to the centerline of the bridges and referenced to benchmarks near the bridges. The cross sections were aligned and positioned using aerial imagery and, therefore, were not accurately geo-referenced to a global horizontal or vertical datum. The cross section at the Glen Avon Bridge was surveyed to the top of the banks. For the other cross sections, only the unvegetated channel was surveyed, and the topographic data for the remainder of the cross sections in the wider flood plain were obtained from the LiDAR data.

The five surveyed cross sections (fig. 3; fig. 13) show a general trend of increasing channel width downstream. At the Glen Avon Bridge, the river is relatively confined within the canyon. At the other cross sections, the river is less confined, with gravel and cobble bars (near the right bank at Feyrer Park, Highway 211, and Highway 213) and low terraces or benches (near the right bank at Highway 211 and Highway 213, and near the left bank at Goods Bridge) within the river banks.

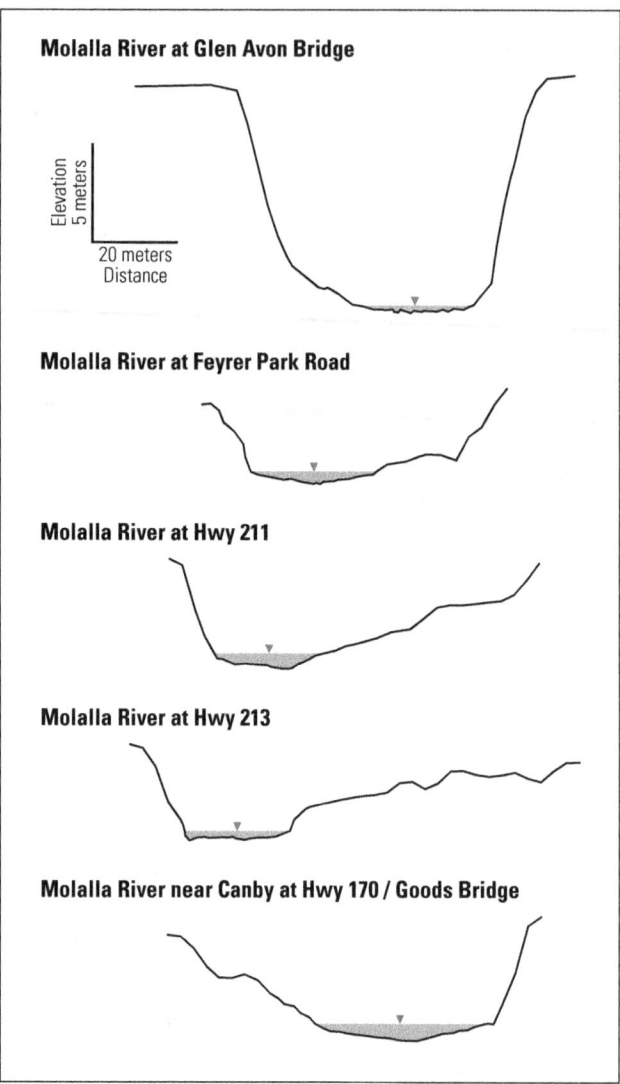

Figure 13. Comparison of measured geometry of five river cross sections surveyed on the Molalla River, Oregon. (The viewer's perspective is looking downstream and the horizontal blue line represents the water-surface elevation on the day of the survey. The vertical exaggeration is 4.6x.)

Analysis of Stage-Discharge Relation at Gaging Station

Following the approach of Klingeman (1973) and Smelser and Schmidt (1998), long-term aggradation trends in the Molalla River were analyzed using stage-discharge relations at the Molalla River near Canby gaging station (14200000). By analyzing the reported stage-discharge relation for the period of record that extends discontinuously back to 1928, the stage for a given discharge was determined. Stage at the gaging station is dependent on downstream hydraulic control and can be used to infer trends in aggradation or incision in the section of river downstream of the gaging station.

For the analysis, four target discharge values were used: 570 m³/s (20,100 ft³/s), which represented the approximate discharge of a 5-year recurrence-interval peak event in the lower Molalla River; 110 m³/s (3,800 ft³/s), the value of discharge exceeded 5 percent of the time; 19 m³/s (670 ft³/s), the value of discharge exceeded 50 percent of the time (median discharge); and 1.7 m³/s (60 ft³/s), the value of discharge exceeded 95 percent of the time. Reported changes in the gaging-site location or the reference datum were noted and used to correct the raw stage data to a common datum.

Figure 14 shows the overall trends in stage in the Molalla River near Canby (14200000) from 1928 to 2010. Each data point represents the reported stage for a given discharge value on that date. Two gaps in the data show when the gaging station was not in operation: November 1958 to October 1963 and November 1977 to October 2000. All data were corrected to accommodate station moves or shifts in the datum such that all the points shown in figure 14 are displayed relative to a common datum. As a result, all stage values for a given discharge are directly comparable, even across data gaps when the gaging station was not operating. From 1928 to about 1948, stage at all analyzed discharge values was relatively constant, shifting by no more than a few centimeters. From 1948 to 1953, stage decreased about 0.25 m for all discharge values, indicating a general trend of incision. This decrease may have been associated with gravel mining commonly practiced along the Molalla River in the middle 20th century (Cole and others, 2004). Between cessation of gaging station operation in 1958 and reactivation of the gaging station in 1963, low-flow stage increased about 0.25 m and high-flow stage decreased a similar amount. From 1958 to 1977, stage decreased at a steady rate. Between 1977 and 2000, the second period when the gaging station was not in operation, stage decreased by another 0.25 m. From 2000 to 2010, stage has fluctuated slightly, but there has been no consistent upward or downward trend. Taken as a whole, the gaging record at Canby shows a general trend of incision for the latter half of the 20th century. More importantly, there was no appreciable aggradation downstream of the gaging station in the past 10 years. Notably, the river downstream of the gaging station did not change appreciably after the December 1964 peak of record.

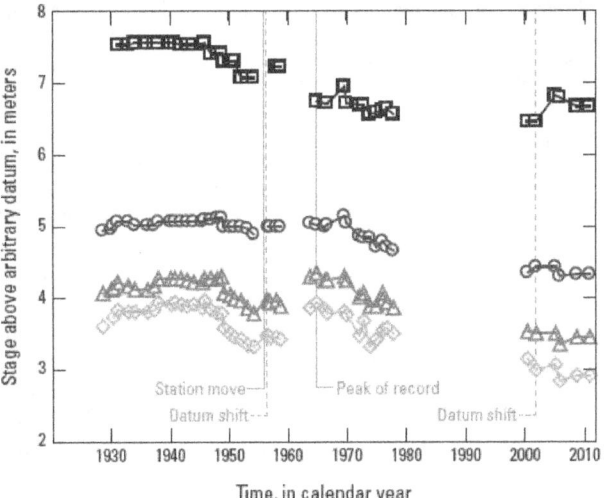

EXPLANATION

Corresponding discharge, in cubic meters per second

☐   570
○   110
△   19
◇   1.7

Figure 14. Trends in stage of water-surface elevation at the U.S. Geological Survey streamflow-gaging station 14200000, Molalla River near Canby, Oregon. (Stage was determined for the 5-year recurrence-interval peak [570 m³/s, or 20,100 ft³/s], the 5 percent exceedance value [110 m³/s, or 3,800 ft³/s], the median discharge [19 m³/s, or 670 ft³/s], and the 95 percent exceedance value [1.7 m³/s, or 60 ft³/s]. Gaging station moves and datum shifts are shown on the graph as is the peak of record on December 22, 1964.)

Spatial Analysis of Flood Plain

Morphologic trends along the Molalla River were determined from historical aerial imagery (table 7). Orthorectified aerial imagery collected in 1994, 2000, 2005, and 2009 was used to digitize channel features and quantitatively analyze changes in the active channel during the last two decades. Note that 1994 imagery for the Molalla River was available as part of the 1995 USGS DOQ imagery set (table 7). Older aerial imagery, from 1936 to 1988, that were not orthorectified allowed for the qualitative analysis of river response over a longer period. Knowledge of the Molalla River flood hydrology between 1994 and 2009 allowed the quantitative analysis of river response to unique peak-flow conditions. For example, the 1994–2000 imagery bounds the large February 2, 1996, high flow that has an estimated discharge of about 900 m³/s (32,000 ft³/s). Similarly, the 2005–2009 imagery bounds a smaller, yet still significant, January 2, 2009, peak event of 691 m³/s (24,400 ft³/s). In contrast, the 2000–2005 imagery bounds a series of modest peak flows, the largest of which was the February 1, 2003, event of 467 m³/s (16,500 ft³/s).

Table 7.   Aerial imagery used in spatial analysis of the Molalla River, Oregon, flood plain.

[**Abbreviations:** BW, black and white; USDA, U.S. Department of Agriculture, USGS, U.S. Geological Survey; DOQ, digital orthophotograph quads; NAIP, National Agriculture Imagery Program]

Year	Type of imagery	Imagery source	Image dates
1936	Aerial photograph (BW)	1936 U.S. Army Corps of Engineers	Apr. 29, 1936
1948	Aerial photograph (BW)	1948 Production and Marketing Administration (PMA), USDA	July 13, 1948
1964	Aerial photograph (BW)	1964 Northern Lights	Sept. 23–24 and 27, 1964
1970	Aerial photograph (BW)	1970 Agricultural Stabilization and Conservation Service (ASCS), USDA	June 25, 1970
1980	Aerial photograph (BW)	1980 Agricultural Stabilization and Conservation Service (ASCS), USDA	April 30 and May 4, 1980
1988	Aerial photograph (BW)	1988 Northern Lights	June 24, July 9, and Sept. 3, 1988
1994	Orthophotograph (BW)	1995 USGS DOQ	June 20, 1994
2000	Orthophotograph (BW)	2000 USGS DOQ	July 24–Aug 5, 2000
2005	Orthophotograph (color)	2005 NAIP	Aug. 3–4, 2005
2009	Orthophotograph (color)	2009 NAIP	June 23, 2009

Methodology of Digitization

Areas of channel migration and flood-plain morphology were digitized using methodology previously developed by the USGS for analysis of the Chetco and Umpqua Rivers. Mapping guidelines and definitions are briefly summarized below; however, a more detailed description of the standard mapping methodology can be found in Wallick and others (2010a; 2010b).

Digitization of channel morphology was completed at a scale of 1:3,000 and only channel features larger than 300 m^2 were digitized. Features smaller than 300 m^2 were integrated into larger, adjacent features. Mapping of morphologic features was primarily confined to the regions adjacent to the active channel, defined as the area typically inundated during annual high flows (Church, 1988). Each digitized data set was reviewed by another team member for quality, accuracy, and consistency.

Base layers used to map geomorphic features within the active channel of the Molalla River include LiDAR-derived imagery (Oregon Department of Geology and Mineral Industries, 2010), digital orthophoto quads from 1994 (1995 DOQ Molalla River imagery) and 2000 (Oregon Geospatial Enterprise Office, 2010); and orthoimagery from 2005 and 2009 (U.S. Department of Agriculture, 2010). For each of the four imagery sets (1994, 2000, 2005, 2009), regions along the river corridor were mapped into one of five general morphologic features: (1) the wetted channel, (2) gravel bars, (3) secondary water features, (4) bedrock outcrops, and (5) flood plain.

For mapped wetted channels, two discrete units were classified. First, the "primary channel" of the Molalla River was mapped on the basis of the visible extent of the wetted perimeter. Then for large tributaries, such as the Pudding River and Milk Creek, the wetted perimeter of the "tributary channel" was mapped roughly 500 m upstream of its confluence with the Molalla River. Next, gravel bars, defined as features greater than 300 m^2 containing exposed bed-material sediment, were delineated and classified as either "flood-plain bar" (adjacent to flood plain) or "island bar" (surrounded by wetted channel). In addition, the amount of vegetation was estimated for all flood-plain-bar and island-bar features as containing bare (less than 10 percent), moderate (10–50 percent), or dense (greater than 50 percent) vegetation cover. Secondary water features, including side channels, backwater sloughs, and disconnected water bodies, were also mapped. Bedrock outcrops, as visible in the aerial imagery, were mapped along the river's active channel. However, only a few outcrops were identified and mapped, which demonstrates the limitation of using aerial imagery to identify bedrock along a vegetated river corridor where subsequent field observations identified extensive bedrock outcrops along most of the corridor. The remaining area that might typically be inundated during annual high flows, based on vegetation distribution and development, was classified as "flood plain," allowing for baseline comparison of morphologic features between different time periods. In addition, the approximate channel centerline of the main-stem Molalla River was digitized for each of the four sets of imagery.

Mapping of channel features was affected by the quality and resolution of available imagery. Some areas of imagery had varying degrees of glare, shadows, or local obstructions of channel features. Errors were introduced by imprecise line placement and orthorectification inaccuracies. To minimize errors and increase the overall precision of the interpretive mapping, all delineated features were regularly checked and quality controlled by other members of the project team.

An important consideration when viewing results from the digitization is that biases in the size of respective morphologic features may exist due to differences in discharge. Although all imagery was acquired during periods of relatively low flow, even a small change in discharge can affect the width of the wetted channel and the area of exposed bars. Discharge data for the Molalla River were not available for all four imagery time periods, so the digitization was not corrected to accommodate for differences in discharge. As a result, no comparative analysis was completed using only the area of the wetted channel.

Channel sinuosity was determined for 1994, 2000, 2005, and 2009 along the main channel of the Molalla River with respect to the geomorphic flood plain. Sinuosity values were determined for each of the geomorphic flood-plain reaches by dividing the reach-segregated channel centerline length by the corresponding geomorphic flood-plain centerline length for that reach.

The active channel, that section of the river corridor relatively free of vegetation that conveys the majority of the water and sediment during high flow, was defined to include the wetted channel, flood-plain and island bars (less than 50 percent vegetation), secondary water features, and bedrock outcrops forming low benches within the channel. Morphologic features that did not contribute to the active channel included flood-plain or island bars with dense (greater than 50 percent) vegetation, flood plain, bedrock forming steep banks, and tributary features. Also excluded from the category of active channel were any secondary water features, island bars, or flood-plain bars that did not form a collective, continuous arc with an upstream and downstream connection to the main active channel. Active-channel width was then determined by dividing the total active-channel area in a given 1-FPkm subreach of the geomorphic flood plain by the length of that subreach. The average active-channel width was determined for each 1-FPkm subreach segment for each of the four years.

Channel centerlines from 1994, 2000, 2005, and 2009 were used to compute average annual migration rates for each of the three time periods between aerial imagery with channel migration averaged over each 1-FPkm subreach segment. The sum migration distance between two sets of aerial imagery was divided by time elapsed between images to determine an average annual migration rate.

The percentage of bank confinement due to bedrock outcrops and revetments, as digitized using observations from the float trips, was quantified for each of the six geomorphic flood-plain reaches. For each geomorphic flood-plain reach, the total length of hardened banks was divided by twice the length of the 2009 channel centerline (to account for the left and right banks separately) in order to determine the percentage of the 2009 channel margin that was confined by bedrock, revetments, or coarse-grained Missoula Flood deposits.

Specific bar area for a given 1-FPkm segment was defined as the total bar area divided by the length along the geomorphic flood-plain centerline. Specific bar area was determined for both the island bars and flood-plain bars, and the three classifications of bar vegetation density (bare, moderate, and dense) were reported along 1-FPkm subreach segments.

Quantitative Trends

Channel sinuosity was highest in GR1, GR3, and GR4, where the channel meandered significantly during the study period (fig. 15). Sinuosity was lowest in GR2, where the channel was relatively straight and ran along the right extent of a wide geomorphic flood plain.

Average active-channel widths were largest and most variable from 1994 to 2009 in GR3 (fig. 16). The smallest and least variable active-channel widths were in GR2 and GR6. The relatively narrow active-channel width in GR2 was due to the numerous confinements, and the narrow active-channel width in GR6 was due to bedrock confining both edges of the channel. GR1 was characterized by relatively large active-channel widths that remained relatively constant through time. In general, active-channel widths in GR3, GR4, and GR5 were variable over time. Indeed, the continuous river corridor from GR5 downstream to GR3 exhibited a large variability in active-channel width from 1994 to 2009 consistent with the pool-riffle morphology, active meandering, and relatively few revetments or confining bedrock walls. Although observed changes in active-channel width of GR1, GR2, and GR6 were largely constant for the four sets of imagery, active-channel width in GR4 and GR5 was largest in the imagery acquired in 2000, after the large February 1996 high flow. By 2005, the active-channel width in GR4 and GR5 had decreased to a value similar to that in 1994. Although the active-channel width in 2009 in GR4 and GR5 did not increase greatly in response to the January 2009 high flow, the active-channel width in GR3 increased significantly.

Consistent with the observed trends in active-channel width, calculated channel-migration rates were largest from GR5 downstream to GR3, and overall channel-migration rates were largest in GR3 (fig. 17), the reach of river prone to extensive management issues for residents living along the banks. In general, channel-migration rates were smallest from 2000 to 2005 throughout the entire study reach, which is indicative of a hydrologic period with smaller peak flows (fig. 5). Channel migration was largest within GR3 for all three periods, with the largest channel-migration rates occurring between 2005 and 2009. From 1994 to 2000, the period bracketing the large February 1996 high flow, channel-migration rates in GR4 and GR5 were generally larger than during the other two periods (fig. 17).

The increased channel-migration rates observed from GR5 downstream to GR3 were due, at least partially, to the presence of fewer channel confinements (fig. 18). The percentage of channel margins confined by bedrock outcrops, revetments, or coarse-grained Missoula Flood deposits varied from 0 percent in GR1, where there were no significant revetments or bedrock outcrops on the banks, to nearly 45 percent in GR6 (fig. 18). In GR2, there were numerous revetments on both the left and right banks. Additionally, the right-bank channel abuts Missoula Flood deposits throughout much of GR2. Although there are revetments in GR3, most

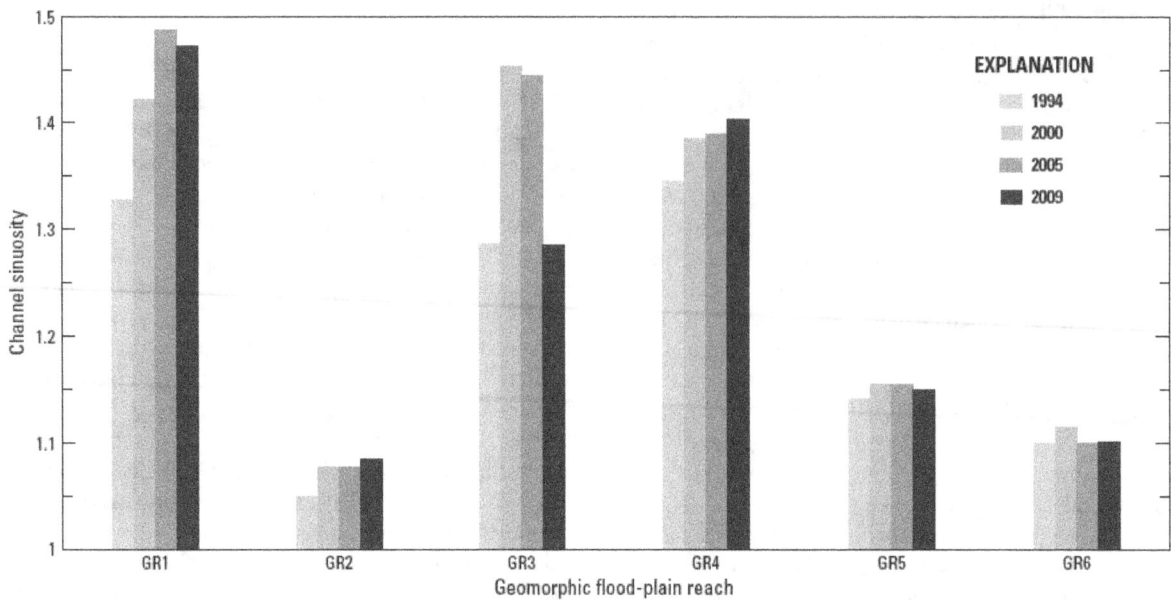

Figure 15.   Historical changes in channel sinuosity, averaged over geomorphic flood-plain reaches, for the Molalla River, Oregon, 1994–2009.

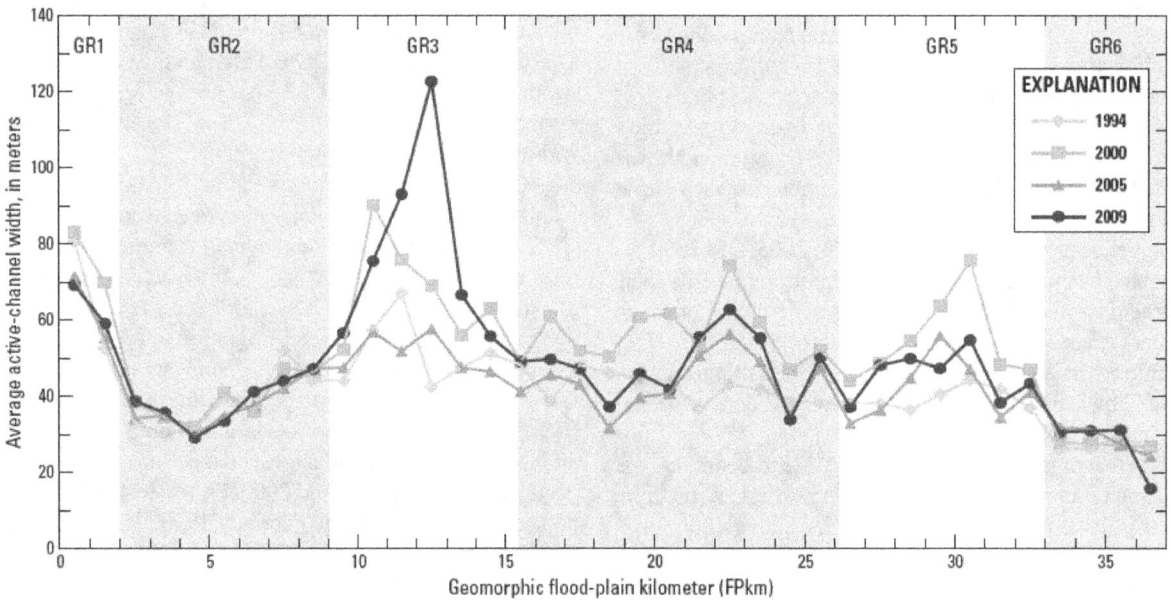

Figure 16.   Historical changes in active-channel width, averaged over each 1-FPkm subreach segment, for the Molalla River, Oregon, 1994–2009.

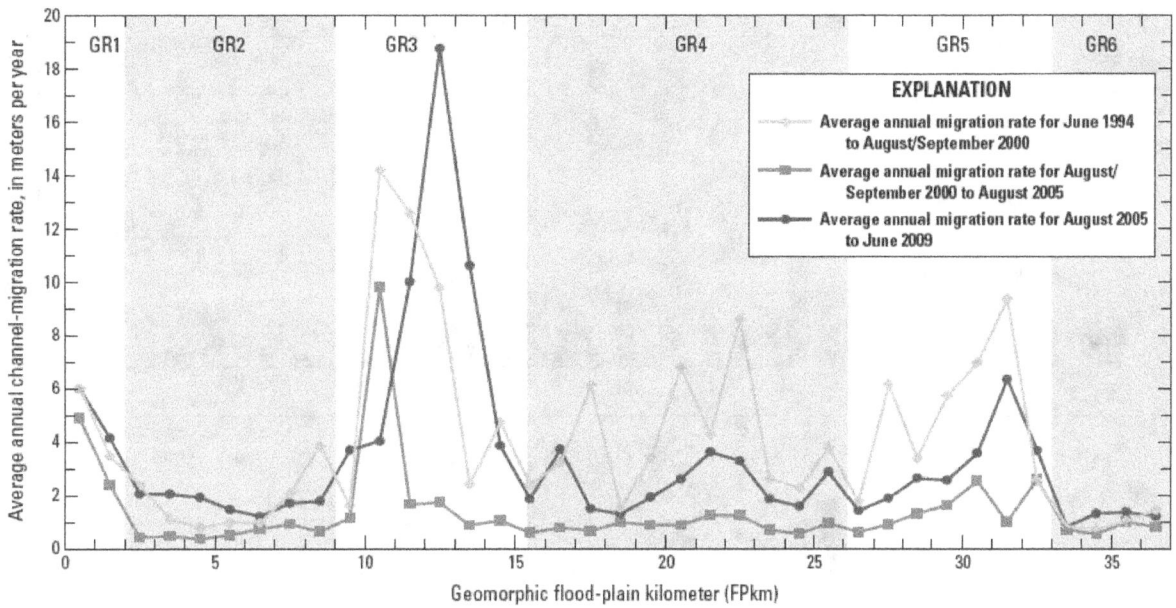

Figure 17. Average annual channel-migration rates, averaged over each 1-FPkm subreach segment, for three time periods between aerial imagery, Molalla River, Oregon, 1994–2009.

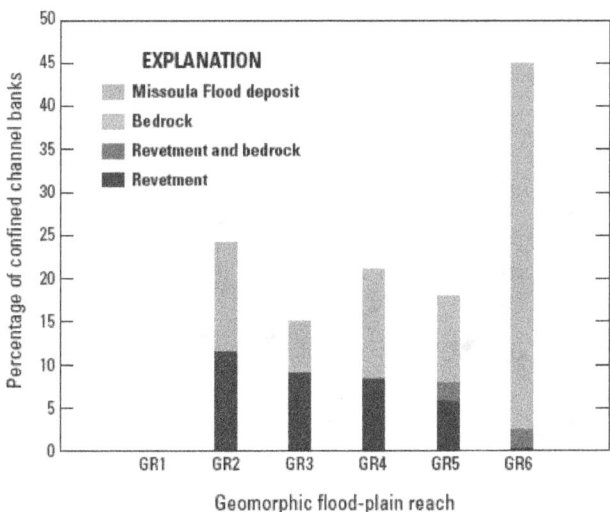

Figure 18. Percentage of channel banks confined by bedrock, revetment, or Missoula Flood deposits along the Molalla River, Oregon, in 2010.

notably along the railroad bridge, along Missoula Flood deposits, and at strategic locations along the left bank, much of the GR3 is free of confinement, which allowed a more active channel between 1994 and 2009. The channel in GR4 is moderately confined by numerous bedrock outcrops along the right bank and by numerous revetments, primarily along the left bank. However, strategic placement of revetments, acting in concert with lateral bedrock outcrops on river right,

has helped to constrain channel movement in GR4. There is a general tendency for the river to remain positioned along prominent lateral bedrock outcrops or along strongly armored revetments with relatively little hydraulic roughness. Qualitative observations of pool depths suggest a spatial correlation between deeper pools and lateral bedrock or well-engineered revetments, which in turn indicates that these erosion-resistant features create hydraulic conditions during large flows that promote effective sediment transport along the bedrock or revetment face, thereby holding the position of the river against the lateral feature over decades. GR5 is confined by a mix of bedrock and revetments on both the left and right banks distributed throughout the reach. GR6 has the highest percentage of confined banks due to bedrock on both the left and right banks, occurring concomitantly on both banks in several sections. Because the revetment data used to generate figure 18 were based on field observations from 2010, the data best describe plan-form conditions of the rivers as determined from the 2009 imagery. Although not evaluated in this study, it is possible that many left-bank revetments, plentiful in GR5 and GR4, were constructed or enhanced following the 1996 flood, which in turn suppressed channel movement in GR5 and GR4 during the 2009 flood.

Specific vegetated bar area (the total height of the columns on the graph) was greatest in GR3, indicative of active channel migration and a large riparian corridor (fig. 19). In contrast, specific bar area was smallest from restricted channel migration in GR2 and GR6 for all four time periods. Specific bar area is variable temporally in portions of GR3, GR4, and GR5. In 1994, most of the flood-plain and island

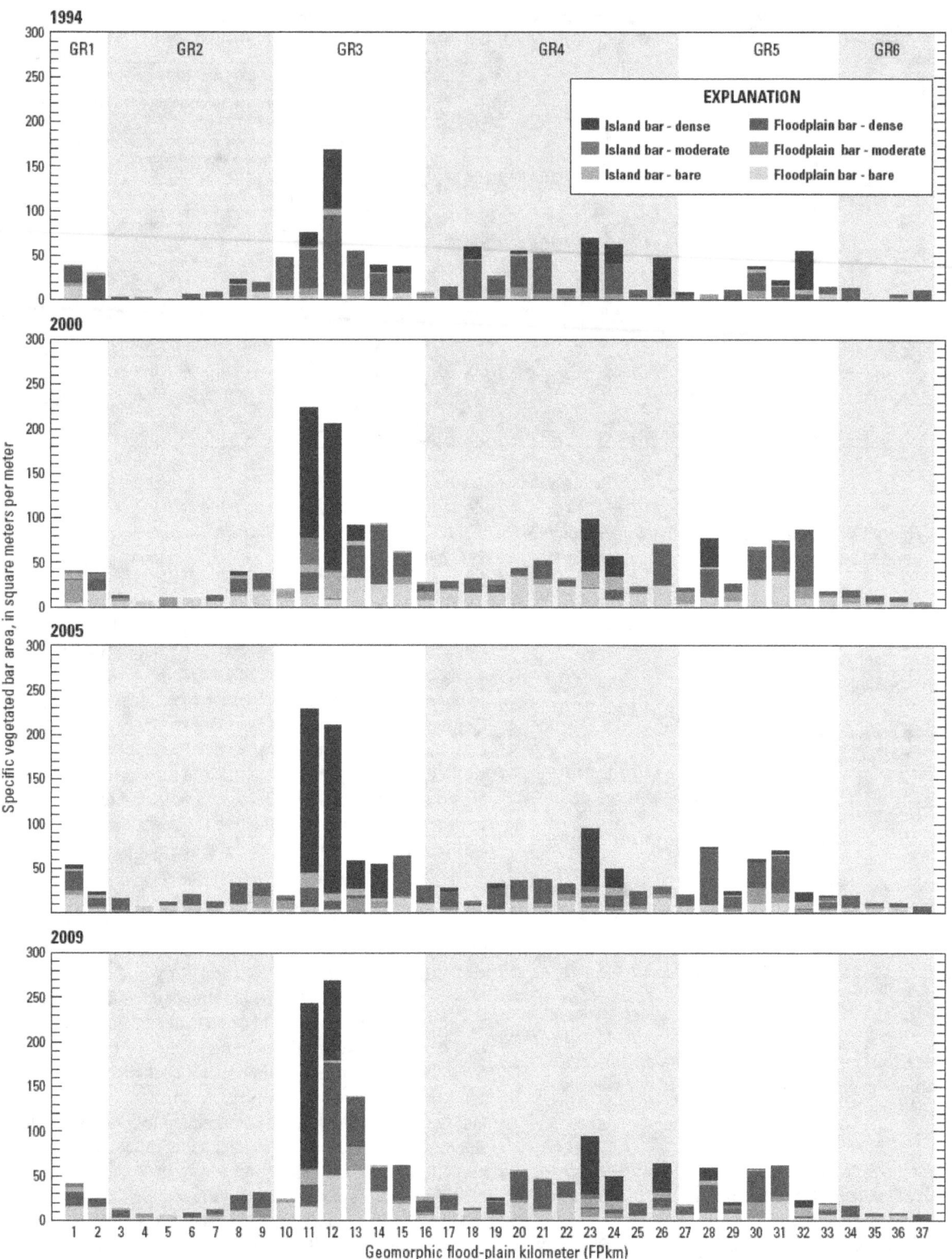

Figure 19.   Historical changes in specific vegetated bar area averaged over each 1-FPkm subreach segment, separated by bar type and vegetation density, along the Molalla River, Oregon, 1994–2009.

bars were densely vegetated. From 1994 to 2000 there was a shift toward a larger proportion of bare flood-plain bars throughout most of the study reach as well as a greater area of island bars in GR3, a river response to the February 1996 high flow. Progressing to 2005, after a period of smaller high flows, the flood-plain bars shifted toward a greater proportion with moderate or dense vegetation coverage as vegetation grew and reclaimed some of the active channel. By 2009, following the January 2009 high flow, some channel-margin vegetation was removed and fresh gravel-bar deposits resulted in a larger proportion of exposed morphologic features.

Qualitative Trends after the 1930s

Geomorphic changes between 1936 and 1988 were qualitatively assessed using historical aerial imagery (table 7). These older images were not rectified and digitized, but general trends of channel width, bar growth, vegetation, and channel migration were analyzed. This qualitative assessment of historical imagery, along with the quantitative analysis from 1994 to 2009, demonstrates the qualitative trends of geomorphic change from 1936 to 2009.

Because problems caused by flooding and channel migration have been most acute in GR3 in recent years, the qualitative analysis was focused here. Figure 20 shows paneled mosaics of aerial imagery for all the available aerial images. In 1936, the Molalla River in GR3 was meandering with established riparian vegetation, extensive gravel point bars, and a wide active channel that appeared to be in a state of continual migration into the adjacent flood plain consistent with the fluvial mechanics of other alluvial rivers in the Willamette River Valley and the greater Pacific Northwest. Qualitatively, the Molalla River of 1936 in GR3 had a geomorphic plan-form appearance similar to the contemporary river (in 2009). The plan-form characteristics of the river in GR3, including sinuosity and active-channel width, changed little from 1936 to 1948, although large areas of exposed gravel bars appeared in the upper extent of GR3 just upstream of the railroad bridge (fig. 20B). These new gravel bars were probably deposited by the relatively large 711 m³/s (25,100 ft³/s, with a recurrence interval of about 10 years; table 4) flow that occurred in January 1948, six months prior to acquisition of the 1948 imagery (table 7).

By autumn 1964 (fig. 20C), the river in GR3 had straightened noticeably as a number of meanders were cut off; however, active-channel width and the extent of vegetation was not greatly different between 1948 and 1964. The imagery collected in 1964 predates the large December 1964 peak of record of 1,240 m³/s (43,600 ft³/s), and channel straightening probably occurred in response to the high flow of November 1960 (fig. 5), estimated to be 980 m³/s (34,500 ft³/s) on the basis of a 7.62 m high-water mark (relative to the gage datum) recorded by the Canby Fire Department (Andy Bryant, National Weather Service, written commun., 2010). The discharge based on this high-water mark was determined using

the current (2011) stage-discharge relation at the Molalla River near Canby gaging station (14200000), but because the trends in stage elevation have been decreasing since 1960 (fig. 14), this discharge estimate is likely too large. The geomorphic effects of the December 1964 peak of record are apparent in figure 20D. Imagery taken in 1970 shows a wide active channel with large, extensive gravel bars free of vegetation in a relatively straight channel of low sinuosity.

By 1980, the active channel was starting to become constricted by encroaching riparian vegetation and meanders were again developing. From 1988 to 1994, a hydrologically quiet period of time with relatively small peak flows (fig. 5), vegetation growth was extensive and the river increased its sinuosity and channel migration. Images acquired in 2000 (fig. 20H), show that the February 1996 high flow again widened the active channel but did not straighten the channel like the 1960 high flow, as evidenced by the 1964 imagery. As discussed earlier, vegetation encroached upon the active channel through 2005 and was subsequently removed by the relatively large January 2009 flood (fig. 5). The active-channel width in the 2009 imagery in GR3 increased, and at least two prominent meanders were cut off (fig.20J). These meander cut offs appeared to supply a large volume of gravel that was deposited locally within 1 km of the cut off, as evidenced by exposed gravel bars in GR3. This locally derived gravel also appears to have caused the marked increase in active-channel width in GR3 near FPkm 12.5 (fig. 16).

Anthropogenic effects on GR3 can also be deduced from the aerial imagery. Although agriculture and timber harvest from the riparian corridor is evident in 1936, neither activity is widespread nor well established, and buildings and other structures are relatively rare. By 1948, agricultural fields started to increase in number and extent on the southern edge of the river, reaching a peak of activity around 1970. From 1970 to 1994, imagery appears to show that agricultural activity waned slightly, but a number of buildings and houses were constructed on the southern edge of the river. It is important to note that the increases in development along the river corridor in GR3 coincided with the relative stasis in hydrologic peak-flow activity from the middle 1970s to 1996. As a result of fewer and less severe floods, the river channel in GR3 between 1975 and 1996 was relatively inactive. It is also worth noting that the railroad bridge at FPkm 14 and Goods Bridge at FPkm 9 have been stable structures since at least 1936, controlling channel movement near each end of GR3. As a result, most channel-migration activity in GR3 has occurred in the reach between the railroad bridge and Goods Bridge.

Downstream in GR2, overall channel-migration activity and geomorphic changes have been relatively modest (compared to that in upstream study reaches) between 1936 and 2009 (fig. 21). These historical trends are consistent with the quantitative trends analyzed for GR2 and can be attributed to significant channel confinement (fig. 18) and a relative lack of gravel accumulation, an observation confirmed with available historical aerial imagery.

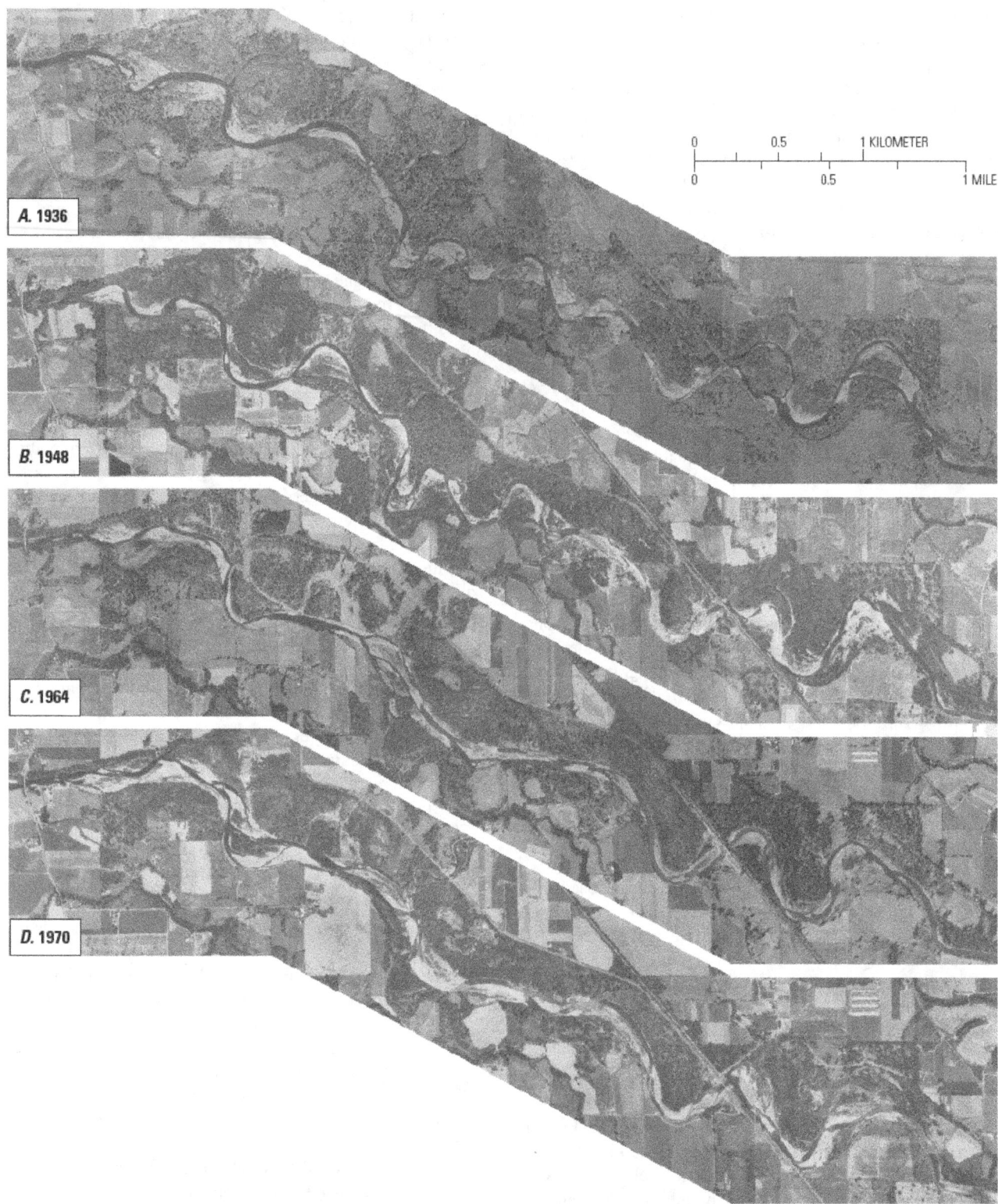

Figure 20.   Paneled mosaics of imagery of the river corridor in geomorphic reach 3 (GR3) for all available sets of aerial imagery. (Panel *K* shows the height-of-water-surface (HAWS) map along GR3 with Goods Bridge and the railroad bridge labeled and the river-stationing coordinates.) An animation of historical geomorphic changes in active channel, at GR3, Molalla River, Oregon is available at http://pubs.usgs.gov/sir/2012/5017/.

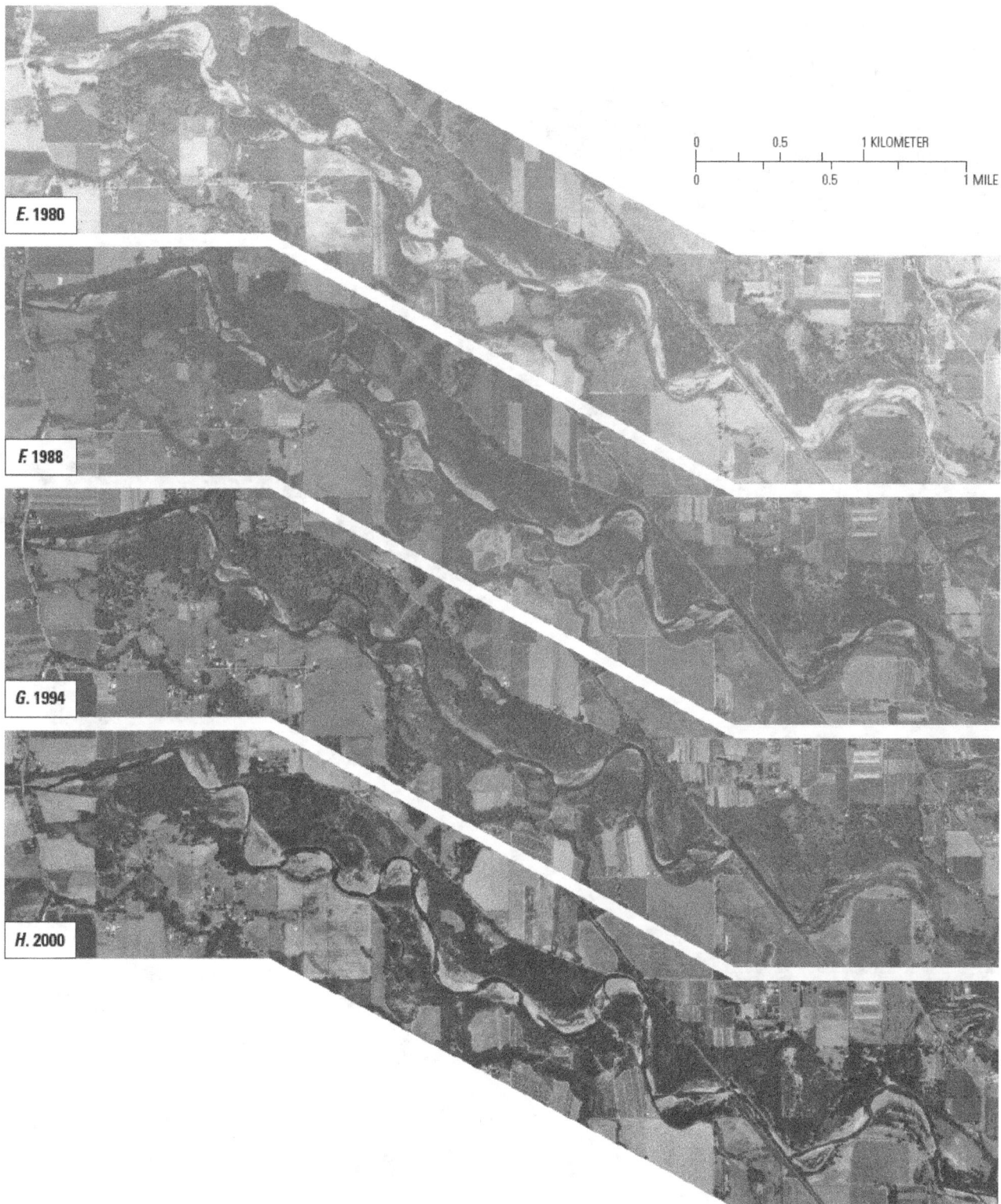

Figure 20. Continued.

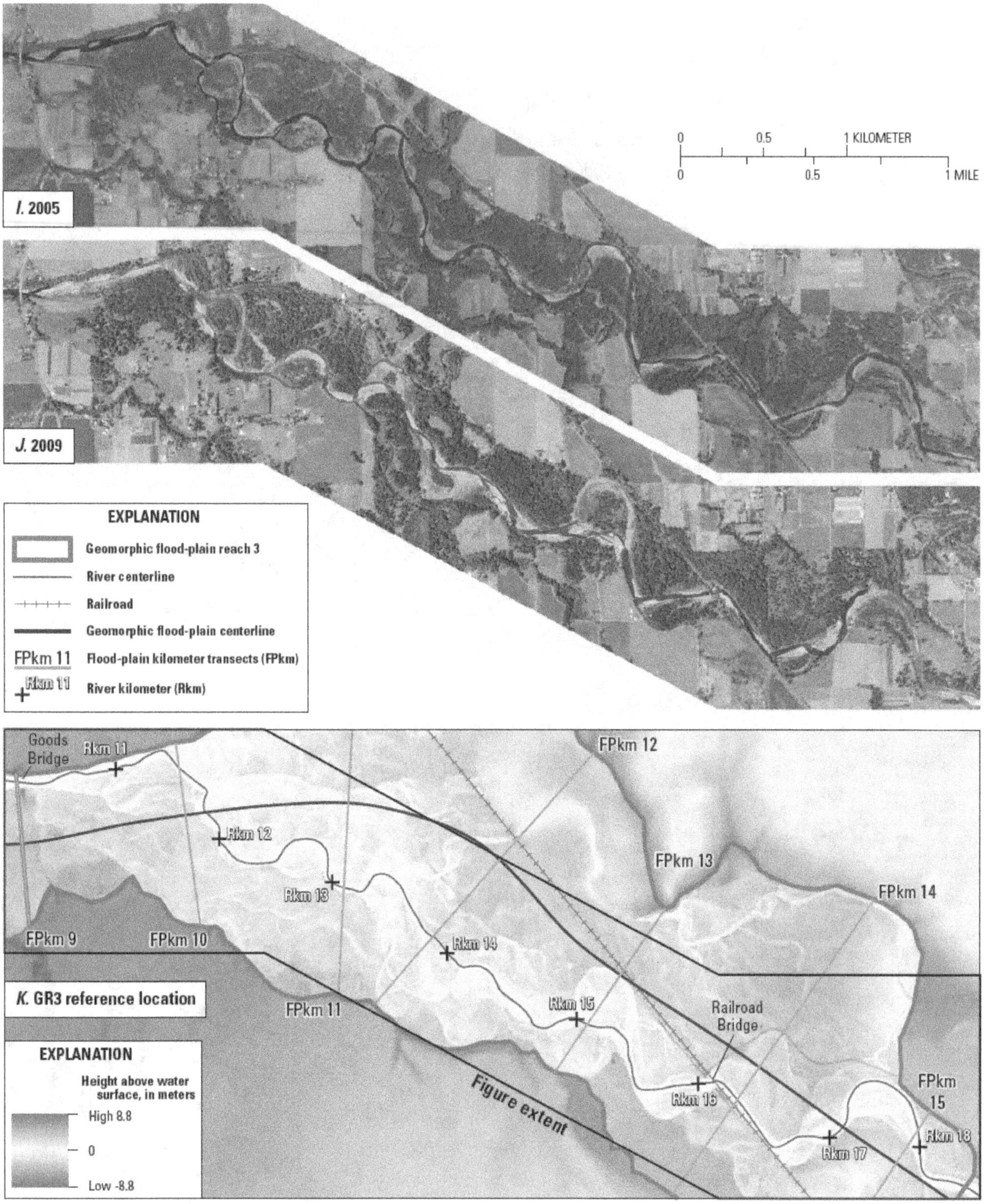

I. 2005

J. 2009

0   0.5   1 KILOMETER

0   0.5   1 MILE

EXPLANATION

Geomorphic flood-plain reach 3

River centerline

Railroad

Geomorphic flood-plain centerline

FPkm 11   Flood-plain kilometer transects (FPkm)

Rkm 11   River kilometer (Rkm)

Goods Bridge   Rkm 11   FPkm 12

Rkm 12

FPkm 13

Rkm 13   FPkm 14

FPkm 9   FPkm 10   Rkm 14

K. GR3 reference location   FPkm 11   Rkm 15   Railroad Bridge

EXPLANATION

Height above water surface, in meters

High 8.8

0

Low -8.8

Figure extent   Rkm 16   FPkm 15

Rkm 17   Rkm 18

Figure 20.   Continued.

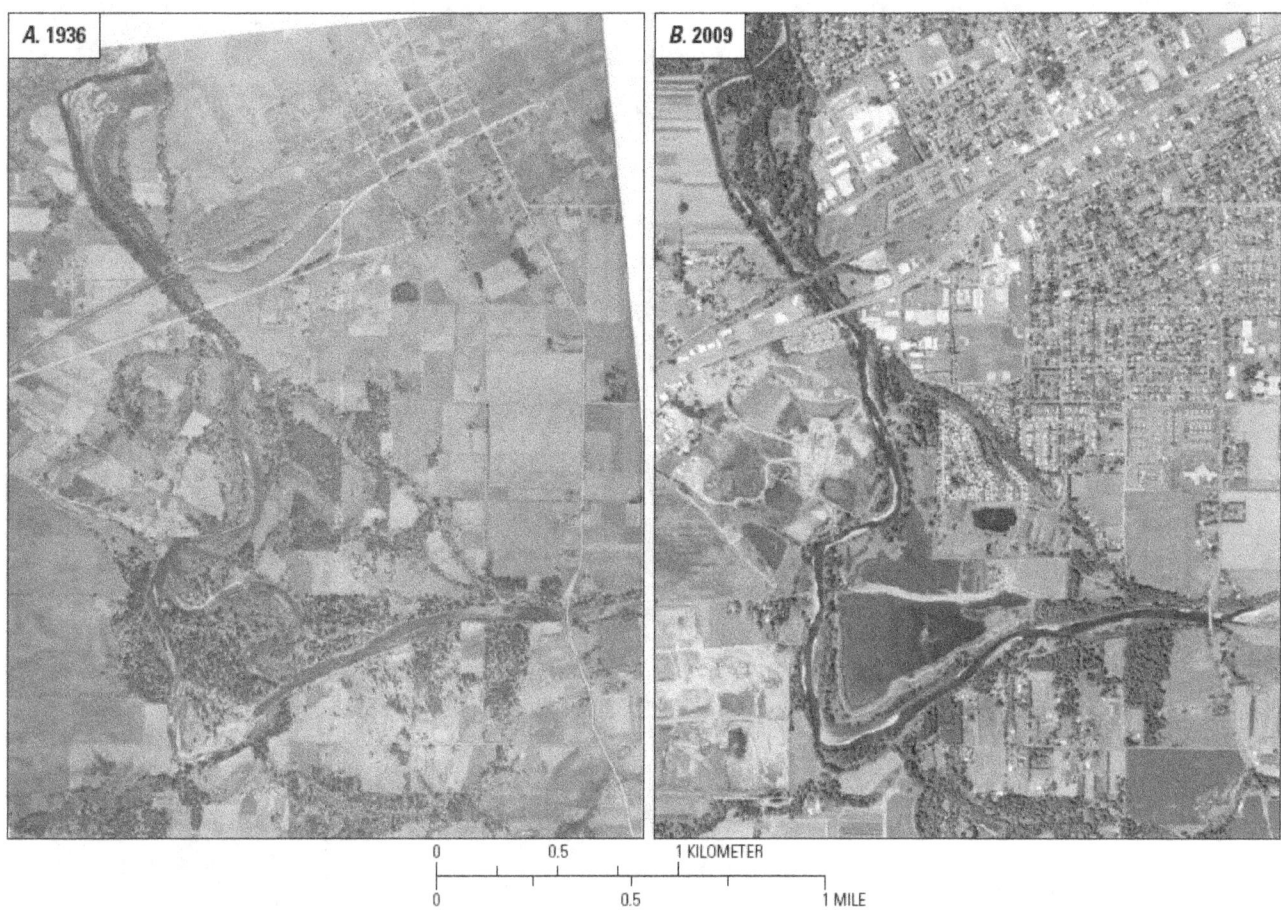

Figure 21. Aerial imagery of the Molalla River in geomorphic reach 2 (GR2) near Canby, Oregon, in (*A*) 1936 and (*B*) 2009, showing the relative stability of channel form and location during most of the 20th century.

Within GR1, the active channel in 2009 was slightly wider than the active channel in 1936, and agriculture has encroached slightly on the outer edges of the wider flood plain (fig. 22). However, the overall geomorphic plan form of the primary river corridor is similar between 1936 and 2009. Riparian vegetation is extensive and well established, and overall sinuosity is similar. One significant change between the dates is that although GR1 in 2009 included the total surface-water flow of the combined Pudding and Molalla Rivers, in 1936 the two rivers were almost separate down to the confluence with the Willamette River (fig. 22A). Although the rivers shared a channel in 1936 for a short distance in the middle of GR1, each river flowed mostly in its own channel and the imagery suggests mixing was negligible.

The 1936 imagery extended no farther upstream than the middle of GR4, but active-channel width, sinuosity, and extent of riparian vegetation in GR4 in 1936 was similar to the geomorphic state of the river in 2009. By 1948, however, GR4 had noticeably changed relative to 1936. The active-channel width was markedly larger, sinuosity had decreased, and a number of avulsions had occurred (fig. 23). Sections of the river corridor in GR4 near FPkm 20 contained extensive bare

gravel bars as wide as 200 m. These gravel bars and the active channel in 1948 were significantly larger than exposed bars observed in any section of the contemporary river corridor. Farther upstream in GR5, active-channel width also appears to have increased by 1948, although the amount of widening was not as severe. Furthermore, the 1948 imagery of GR6, still farther upstream, showed neither a widening of the active channel nor extensive gravel deposits. The 1948 imagery, however, does show aggressive logging and dendritic road networks that were presumably used to access forested lands in the catchment away from the river. Although a large high flow could, in theory, cause the geomorphic response observed in the 1948 imagery of GR4, neither the gaging station record (fig. 5) nor the 1948 imagery from GR1, GR2, GR3 (fig. 20), or GR6 support such a hypothesis. Taken in context, it appears that active logging in the upper Molalla River catchment mobilized substantial amounts of sediment that exceeded the transport capacity of the channel in GR4, resulting in a significantly wider active channel and extensive gravel-bar deposits in 1948. Subsequent imagery in 1964 suggests this pulse of sediment both moved downstream and was extracted by gravel-mining operations that were apparent in the imagery.

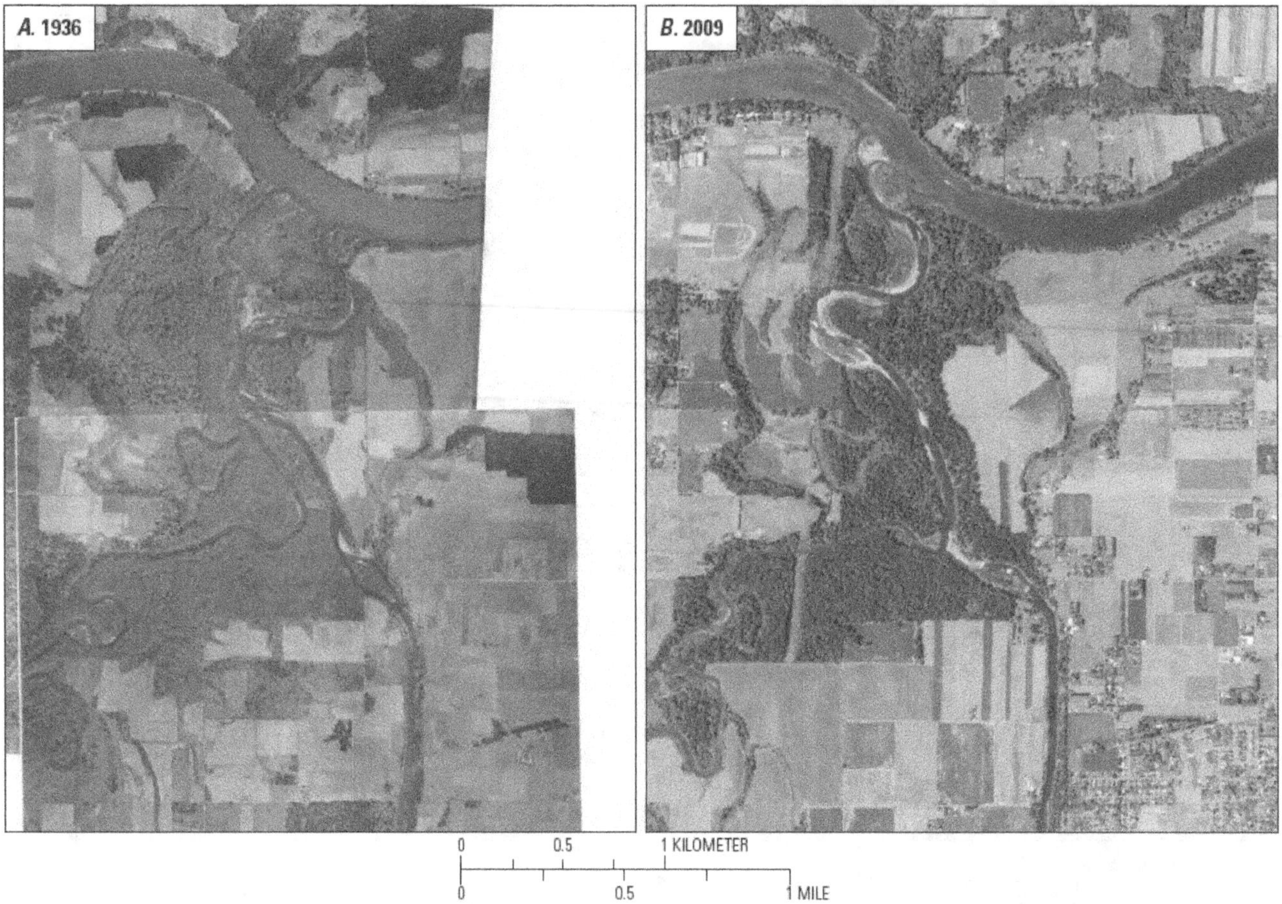

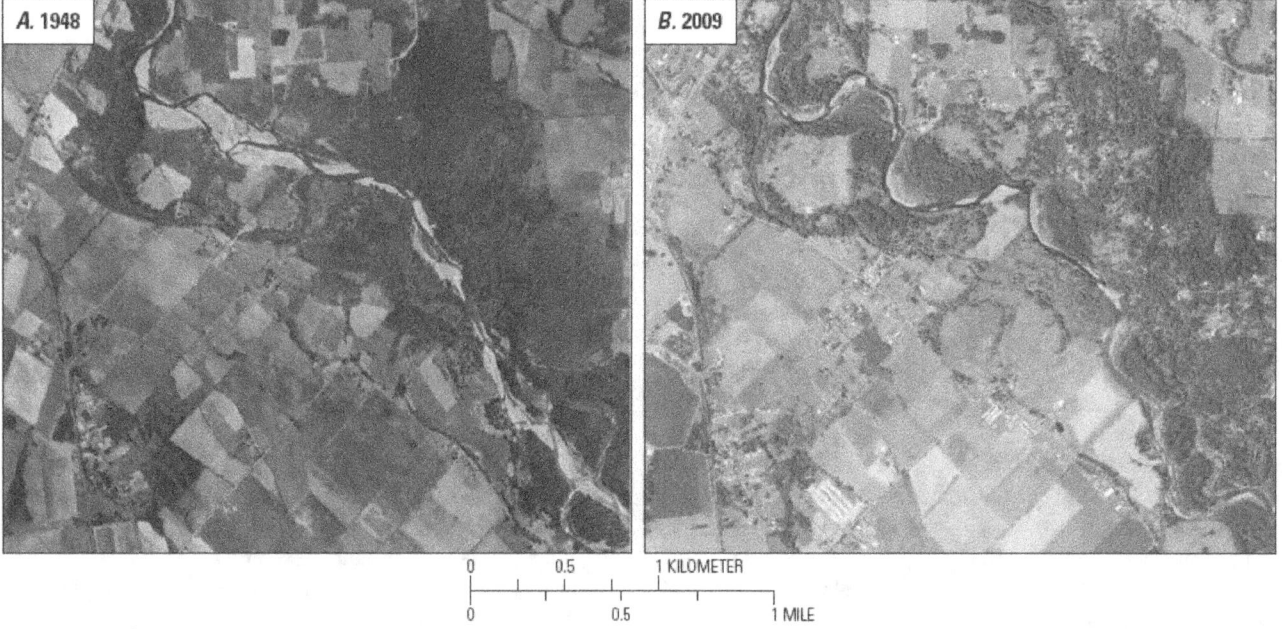

Figure 22.   Aerial imagery of the Molalla River in geomorphic reach 1 (GR1) from near the confluence of the Willamette River in (*A*) 1936 and (*B*) 2009, showing a wide and active flood plain between the Pudding River and the Molalla River, Oregon.

Figure 23.   Aerial imagery of the Molalla River, Oregon, in geomorphic reach 4 (GR4) just upstream of Highway 213 bridge (FPkm 19.5) to about FPkm 23 in (*A*) 1948 and (*B*) 2009. (Some hydrologic high-flow event, a pulse of sediment and aggradation, or a combination of both led to a wide active channel and expansive gravel bars in 1948; subsequent revegetation decreased the active-channel width and increased sinuosity by 2009.)

In GR5, the 1964 imagery shows that the high flow of November 1960 widened the active channel but did not lead to major avulsions. The channel response in GR6 that is visible in the 1964 imagery was not significant. By 1970, however, imagery shows that there was widespread geomorphic change from GR6 downstream through GR4. Vegetation was removed throughout the upper river corridor, and new and reworked gravel bars appeared in GR5 and GR4. The 1970 imagery of GR6, which showed a removal of vegetation, did not show widespread deposits of new sediment. By 1980, riparian vegetation had begun to constrict the active channel throughout GR4, GR5, and GR6, and this process of revegetation and suppressed channel migration continued through the 1994 imagery set, resulting from the relatively quiet hydrologic period from 1975 to 1996.

Channel Response to Flooding

Inferences about the response of the Molalla River to the magnitude of high flows recorded in the gaging station record were made by comparing aerial imagery acquired from 1936 to 2009. As with other lowland rivers in western Oregon and Washington (for example, O'Connor and others, 2003; Beechie and others, 2006), the Molalla River is subject to a flashy peak-flow regime that acts to promote a relatively wide active channel, large channel migration rates, and frequent avulsions. In turn, vegetation growth works to reduce the active-channel width. In the Molalla River, geomorphic response is dependent on the size of the high flows as well as on anthropogenic constraints, revetments, and the underlying controls of the specific geomorphic reach. Pulses of sediment, whether generated from the catchment and transported through the river corridor or derived locally from channel migration and avulsions, can also drive geomorphic change in some sections of the river. The most upstream reach, GR6, appears to be relatively insensitive to the high flows or pulses of sediment. Although the large recorded high flow of December 1964 widened the active channel slightly within GR6, changes in plan-form position were modest. GR6 is also largely a transport reach as little sediment seems to gather. In contrast, the relatively alluvial reaches of GR5 down through GR3 are sensitive to both high flows and pulses of sediment. Wide gravel bars, active channel migration, and occasional avulsions have been common in GR5, GR4, and GR3 since 1936. In the absence of large flows, vegetation encroachment acts to reduce active-channel width and increase sinuosity. Revetments, however, more effectively stabilize the channel plan form. Although anecdotal accounts of active channel movement in GR3 have suggested increased channel mobility in the past 10–20 years, historical aerial imagery shows that channel mobility of the Molalla River in GR3 is likely no greater in the 21st century than it has been historically back to 1936. Relative to GR5 and GR4, the larger quantitative rates of channel migration and active-channel width in GR3 is a function of the few revetments and sparse bedrock control. Farther downstream, GR2 has been restricted by revetments and channel bedrock that have limited channel mobility relative to that in other reaches. It also appears that GR2 has not been subject to widespread sedimentation. GR1 is a reach subject to gradual channel migration across the geomorphic flood plain that reflects the fluvial processes of the pre-development river.

Synthesis of Geomorphic Analyses

The geomorphic analyses made in this study show that, in contrast to early hypotheses suggesting widespread sedimentation, it is unlikely the reach of the Molalla River in GR3 has been aggrading in the past ten years. Observations of the channel bedrock just upstream and downstream of GR3 (fig. 11) suggest the convexity in the long profile observed in GR3 is not the result of a wedge of alluvium. Although the total depth of alluvium above bedrock is unknown, the water-surface profile suggests that it is likely the Molalla River here has adjusted its gradient to transport sediment through this reach. The supply of bedload entering GR3 from upstream is probably small. Moreover, gaging-station data dating back to 1928 show that at the location of the Canby gaging station, the river has incised about 0.5 m since the 1960s (fig. 14). Aerial imagery acquired from 2005 to 2009 (fig. 20) also shows that the likely source of sediment that has affected some sections of the river in GR3 was derived locally from meander cut offs and not from upstream. An example of such a gravel deposit from GR5 is shown in photograph 3, p. 44.

This study also showed qualitatively that channel-migration rates in GR3, which can be an indicator of increased sedimentation, were not significantly greater in the past 10 years than those rates observed during the 20th century. The avulsions and stranded channels observed recently have occurred repeatedly in the past. It is also important to note that although high flow events from 1996 to 2009 were greater than those in the previous two decades (1975–95), these events were not unusually large if evaluated in the larger hydrologic context of the Molalla River. For example, the recurrence interval for the February 1996 high flow was between 25 and 50 years (table 4). Similarly, the January 2009 high flow was about a 10-year event. Both of these events were large, but neither was exceptionally so.

Photograph 3.   Gravel deposit downstream of a channel avulsion in GR5. (Photograph taken by Kurt Carpenter, July 2010.)

There is little evidence that the river in GR6, GR5, or GR4 is aggrading or otherwise filling with sediment. Channel bedrock is present through the river corridor in these reaches (fig. 11), and channel migration rates between 2005 and 2009 (fig. 17) were small compared to migration rates observed in GR3. Although there were indications that a pulse of sediment accumulated in GR4 by 1948, no contributing effects of this sediment are evident in the channel dynamics of the last few decades. Upon closer inspection, the gravel and cobble bars upstream of the Highway 211 bridge in GR5 and GR6 that appeared elevated when viewed from the flood plain were found to be deposited in the river corridor at an elevation controlled by underlying bedrock, not by excess sediment. Although not directly evaluated or calculated, the sediment transport capacity of the Molalla River appeared to exceed the available sediment load. Moreover, there is little evidence to support the hypothesis that large volumes of sediment are being transported into the study reach from the upper part of the catchment.

Streamflow, Water Quality, and Algal Conditions

Data collection for this 2010 study builds on previous water-quality investigations by the USGS in 2000 in the main-stem Molalla River (U.S. Geological Survey, unpub. data). Water samples were collected, field parameters were measured, and benthic community conditions were assessed twice at five locations in the lower Molalla River during summer low-flow conditions in 2010 (fig. 3; table 2). The water samples were analyzed to determine the amount of nutrients available to support algal growth, and field parameters such as dissolved oxygen and pH were measured to gage the relative effects of algal photosynthesis and benthic respiration on water quality. Excessive algal production can produce high (alkaline) pH and result in low concentrations of dissolved oxygen that can be harmful for aquatic life including fish. Sampling of benthic algae (periphyton) was conducted to provide information on the specific types of algae in the riffles and chlorophyll-*a* was used to estimate algal biomass levels. Qualitative surveys of benthic invertebrates were conducted to gage the health and quality of these organisms. Many invertebrates consume algae, and these secondary producers are important food resources for fish including salmon and steelhead.

A cool and wet spring in 2010 delayed and shortened the algal growing season, which may have limited the accumulation of biomass in the river and dampened algal-photosynthesis induced fluctuations in pH and dissolved oxygen levels. Streamflow at the time of the water quality and algal sampling in 2010 ranged from 2.5 to 3.6 m³/s (88 to 127 ft³/s), which was 12–18 percent higher in July and August and 22–28 percent lower in September compared with the long-term average flows at the USGS gaging station near Canby. Streamflow has a direct effect upon water quality and algal conditions by affecting the water temperature, dilution rate, and resulting concentrations of nutrients and other solutes. Also, streamflow and water depth affect light available for algae to photosynthesize, and the amount of light reaching the streambed dictates, in part, the areal extent of algal growth. Vast areas of shallow riffles appear during periods of low flow in summer that provide ideal habitat for periphyton. In some wider sections of the stream, periphytic growth occurs across the entire channel and in other areas it is concentrated primarily along the channel margins.

Field Parameters

Field parameter data, including water temperature, dissolved oxygen (DO), pH, specific conductance, chlorophyll-*a* (water column), and turbidity were collected at each of the five water-quality sites using a calibrated Yellow Springs Instruments 6600 EDS multi-parameter sonde. Water temperature, pH, and DO measurements were made in the main flow in the early morning and again in late afternoon to characterize diel minimum and maximum values during sunny and warm periods in August and September 2010 (table 8).

The Molalla River exhibited the typical downstream trend of increasing water temperature and specific conductance (fig. 24) consistent with the "river continuum" of natural changes associated with downstream increases in channel width, reduction in of the amount of riparian shading, and other downstream changes that occur in many rivers (Vannote and others, 1980). The increase in specific conductance at the two downstream sites was most likely due to a greater amount of agricultural and urban influences (Cole and others, 2004) along with inputs in the lower river from tributary drainages with similar land uses and possibly inflows of groundwater.

Water temperatures continued to be high in the Molalla River during summer 2010. The early August afternoon water temperature at Glen Avon Bridge (Rkm 42.8) was 19.5°C and higher (22–24°C) at the downstream sites in 2010 (fig. 24; table 8). Temperatures were not as high in 2010 as they were in 2004 when ODEQ conducted a basin-wide assessment of water temperatures in the Molalla River that led to the development of a TMDL analysis (Williams and Bloom, 2008). A water-temperature survey of the Molalla River conducted in 2001 (U.S. Geological Survey, unpub. data) shows that the peak temperatures steadily increased from the Table Rock Fork to the Highway 213 Bridge with only a modest increase in water temperature between the Highway 213 and Knights bridges during this low flow water year (fig. 25). Inputs of cooler water in the lower reach, although not investigated during this study, are possible and could have helped offset further warming downstream at Knights Bridge.

Morning and afternoon values of DO and pH for 2000 and 2010 are shown in figure 26. Although the water temperatures exhibited an increase downstream of the Glen Avon Bridge, minimum diel DO and pH declined downstream in 2010. The maximum (afternoon) DO showed modest downstream increases owing to algal photosynthesis, whereas the patterns for maximum diel pH were not consistent, with observed increases between Highway 213 Bridge (Rkm 39) and Knights Bridge (Rkm 6.9) sites in 2010, but not in 2000 (fig. 26).

In 2010, the minimum dissolved oxygen concentrations ranged from 7.6 to 9.8 mg/L, or 84.3 to 97.8 percent saturation, with the lower values occurring at the downstream sites during the morning hours (fig. 26). Longitudinal patterns in diel variations of dissolved oxygen and pH show a greater influence from algal photosynthesis in the lower river, particularly at the Goods and Knights Bridges sites. The maximum DO concentration (11.1 mg/L at Knights Bridge in September 2010) correspond to 122 percent of saturation (supersaturated conditions), which resulted from algal photosynthesis.

Table 8. Diel field parameter data and diel ranges in the main-stem Molalla River, Oregon, August–September 2010.

[Shading indicates diel ranges. **Abbreviations:** ft³/s, cubic foot per second; C, degrees Celsius; µS/cm, microsiemens per centimeter; mg/L, milligram per liter; µg/L, microgram per liter; <, less than; –, no data]

Site	River kilometer	Date	Time	Streamflow (ft³/s)	Water temperature (°C)	Specific conductance (µS/cm)	Dissolved oxygen (mg/L)	Dissolved oxygen in percent saturation	pH, in standard units	Turbidity, in formazin nephelometric units	Chlorophyll-a (µg/L)
Molalla River upstream of Glen Avon Bridge	42.8	Aug. 2, 2010	1600	–	19.5	52	9.4	103.3	8.0	0.7	0.7
Molalla River upstream of Highway 211 Bridge	30.1	Aug. 2, 2010	1620	–	22.2	54	9.1	106.1	7.9	0.5	1.0
Molalla River upstream of Highway 213 Bridge	23.5	Aug. 2, 2010	1640	–	22.2	58	9.4	109.1	7.6	0.9	0.8
Molalla River upstream of Goods Bridge	9.7	Aug. 2, 2010	1710	126	23.5	66	9.4	112.4	7.9	0.7	0.9
Molalla River downstream of Knights Bridge	4.3	Aug. 2, 2010	1740	–	24.1	73	9.8	118.9	8.3	0.7	1.3
Molalla River upstream of Glen Avon Bridge	42.8	Aug. 3, 2010	0540	–	16.4	53	9.3	96.5	7.7	1.3	0.7
Molalla River upstream of Highway 211 Bridge	30.1	Aug. 3, 2010	0610	–	17.3	54	8.8	92.8	7.5	1.0	0.9
Molalla River upstream of Highway 213 Bridge	23.5	Aug. 3, 2010	0630	–	18.9	59	8.1	88.3	7.4	1.0	1.2
Molalla River upstream of Goods Bridge	9.7	Aug. 3, 2010	0650	124	19.8	67	7.6	84.3	7.3	1.0	1.7
Molalla River downstream of Knights Bridge	4.3	Aug. 3, 2010	0720	–	20.0	77	7.9	87.5	7.4	1.4	1.0
Molalla River upstream of Glen Avon Bridge	42.8	Aug. 2–3, 2010	Diel range	–	3.1	1.0	0.0	6.8	0.3	–	–
Molalla River upstream of Highway 211 Bridge	30.1	Aug. 2–3, 2010	Diel range	–	4.9	0.0	.3	13.3	0.3	–	–
Molalla River upstream of Highway 213 Bridge	23.5	Aug. 2–3, 2010	Diel range	–	3.3	1.0	1.3	20.8	0.2	–	–
Molalla River upstream of Goods Bridge	9.7	Aug. 2–3, 2010	Diel range	–	3.7	1.0	1.8	28.1	0.6	–	–
Molalla River downstream of Knights Bridge	4.3	Aug. 2–3, 2010	Diel range	–	4.1	4.0	2.0	31.4	0.9	–	–
Molalla River upstream of Glen Avon Bridge	42.8	Sept. 14, 2010	1400	–	15.0	54	10.5	106.1	8.0	<0.5	1.1
Molalla River upstream of Highway 211 Bridge	30.1	Sept. 14, 2010	1430	–	17.3	55	10.1	106.6	7.8	<0.5	0.8
Molalla River upstream of Highway 213 Bridge	23.5	Sept. 14, 2010	1520	–	18.7	61	10.3	112.2	7.6	<0.5	0.1

Table 8. Diel field parameter data and diel ranges in the main-stem Molalla River, Oregon, August–September 2010.—Continued

[Shading indicates diel ranges. **Abbreviations:** ft³/s, cubic foot per second; C, degrees Celsius; µS/cm, microsiemens per centimeter; mg/L, milligram per liter; µg/L, microgram per liter; <, less than; –, no data]

Site	River kilometer	Date	Time	Streamflow (ft³/s)	Water temperature (°C)	Specific conductance (µS/cm)	Dissolved oxygen (mg/L)	Dissolved oxygen in percent saturation	pH, in standard units	Turbidity, in formazin nephelometric units	Chlorophyll-*a* (µg/L)
Molalla River upstream of Goods Bridge	9.7	Sept. 14, 2010	1550	89	19.5	68	10.4	114.2	7.8	<0.5	2.1
Molalla River downstream of Knights Bridge	4.3	Sept. 14, 2010	1650	–	19.8	77	11.1	122.0	8.4	<0.5	0.8
Molalla River upstream of Glen Avon Bridge	42.8	Sept. 15, 2010	0650	–	14.4	56	9.8	97.8	7.5	<0.5	0.9
Molalla River upstream of Highway 211 Bridge	30.1	Sept. 15, 2010	0720	–	15.7	57	9.1	92.7	7.4	0.5	1.0
Molalla River upstream of Highway 213 Bridge	23.5	Sept. 15, 2010	0740	–	16.3	62	8.7	89.7	7.2	<0.5	0.7
Molalla River upstream of Goods Bridge	9.7	Sept. 15, 2010	0810	89	16.9	70	8.4	87.2	7.2	<0.5	0.8
Molalla River downstream of Knights Bridge	4.3	Sept. 15, 2010	0830	–	17.2	96	8.6	89.8	7.4	0.5	1.1
Molalla River upstream of Glen Avon Bridge	42.8	Sept. 14–15, 2010	Diel range	–	0.6	2.0	0.8	8.3	0.5	–	–
Molalla River upstream of Highway 211 Bridge	30.1	Sept. 14–15, 2010	Diel range	–	1.6	2.0	1.0	13.9	0.4	–	–
Molalla River upstream of Highway 213 Bridge	23.5	Sept. 14–15, 2010	Diel range	–	2.4	1.0	1.7	22.5	0.4	–	–
Molalla River upstream of Goods Bridge	9.7	Sept. 14–15, 2010	Diel range	–	2.6	2.0	2.0	27.0	0.6	–	–
Molalla River downstream of Knights Bridge	4.3	Sept. 14–15, 2010	Diel range	–	2.6	19.0	2.5	32.2	1.0	–	–

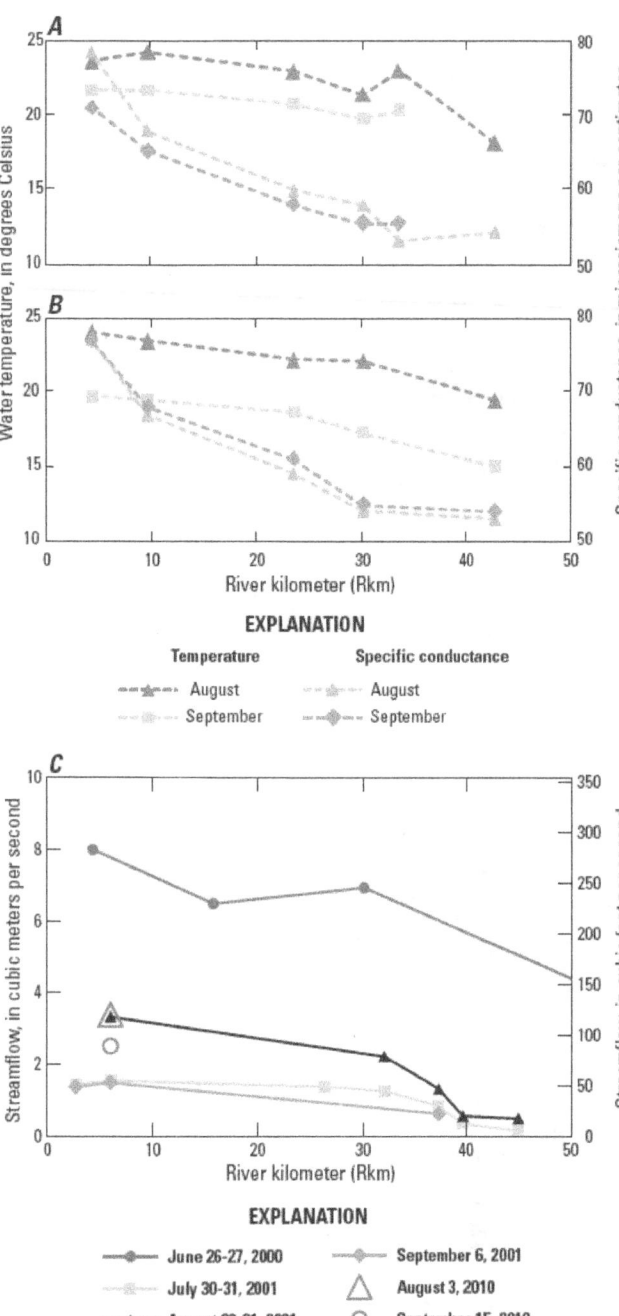

Figure 24.   Afternoon water temperature and specific conductance values for (*A*) August–September 2000, (*B*) August–September 2010, and (*C*) longitudinal streamflow for various periods in the Molalla River, Oregon.

Diel swings in pH, with maximum values up 8.3 and 8.4 units during the afternoon and lower values in the morning, are synchronized with dissolved oxygen swings caused by daily cycles of photosynthesis and respiration. The observed daily ranges in pH were not, however, as great as those in the Clackamas River to the north, where pH fluctuations caused by algal photosynthesis can exceed 1–2 units in a day (U.S. Geological Survey, 2011). Daily ranges in pH were closer to 0.5–1 units in a day during USGS samplings in 2010, and daytime maximum pH values were always within the State of Oregon standard of 6.5–8.5 units.

Periodic measurements of pH were taken by ODEQ in the lower Molalla River since 1965 at Highway 99E and more recently at Knights Bridge (fig. 27). Although pH has been as high as 8.5 units, the highest value in 2010 was lower—only 8.0 units—on August 12 (Oregon Department of Environmental Quality, 2010). The lower maximum pH value could have resulted from the cooler than average temperatures and higher flows that delayed the algal growing season in 2010. Note that although pH values in figure 27 appear to be higher in recent years, that the time of day measurements are taken has a large effect, as would be expected in a productive river showing diel fluctuations in pH and dissolved oxygen. The highest values of pH occur after 2 P.M.

Big steelhead in the upper Molalla River. (Photograph taken by Rob Russell, 2008.)

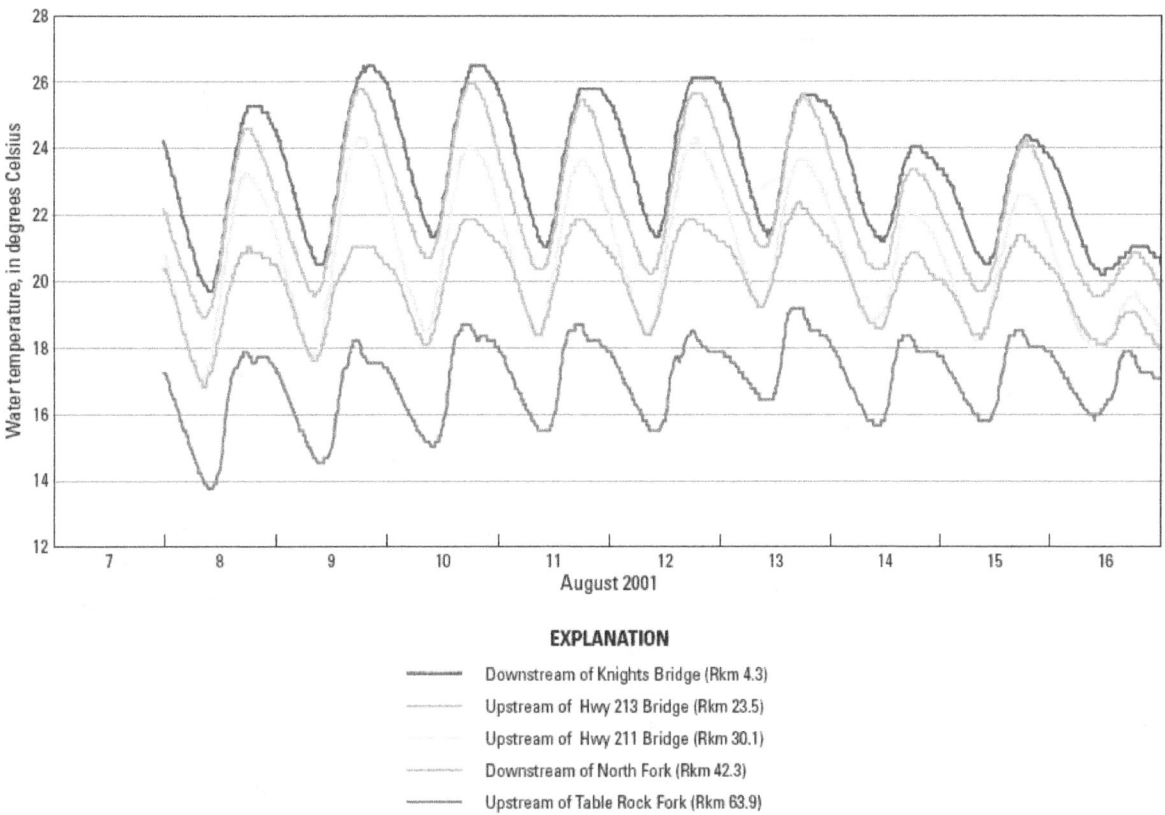

Figure 25. Water temperatures in the Molalla River, Oregon, August 2001.

The water-quality standards in effect for the main-stem Molalla River for water temperature and dissolved oxygen are dictated by the fish-use designation and whether salmon or steelhead are rearing and migrating, or spawning. The middle and upper Molalla River (upstream of the Highway 211 Bridge) and the North Fork Molalla River are currently designated as core cold water habitat for fish, and the lower main stem provides rearing and migration habitat for salmon and trout (Oregon Department of Environmental Quality, 2003). Spawning maps from Oregon Department of Environmental Quality (2005) indicate that salmon or steelhead spawn in the lower river (downstream of the Hwy 211 Bridge) from October 15 to May 15; in the short reach just upstream of the Hwy 211 Bridge to near the confluence with Dickey Creek from September 1 to June 15; and in the uppermost reach, which extends from near Dickey Creek to the headwaters, from August 15 to June 15. On the basis of this information, and relative to data collected for this study, the fish spawning criteria apply only to the Glen Avon Bridge site on September 15, 2010. At this time, the water temperature, pH, and dissolved oxygen values appear to meet water-quality standards although temperature and oxygen data were collected on just one day whereas the State standards are based on 7- and 30-day average maximum and minimum values, respectively. The remaining data collected for this study were evaluated against the water-quality standards in effect during non-spawning periods. In this case, only dissolved oxygen at Goods and Knights Bridges (7.6 and 7.9 mg/L) did not meet the 8 mg/L criteria in the early morning, but again, these were one-day measurements targeting daily minimum values, not 30-day average minimum concentrations.

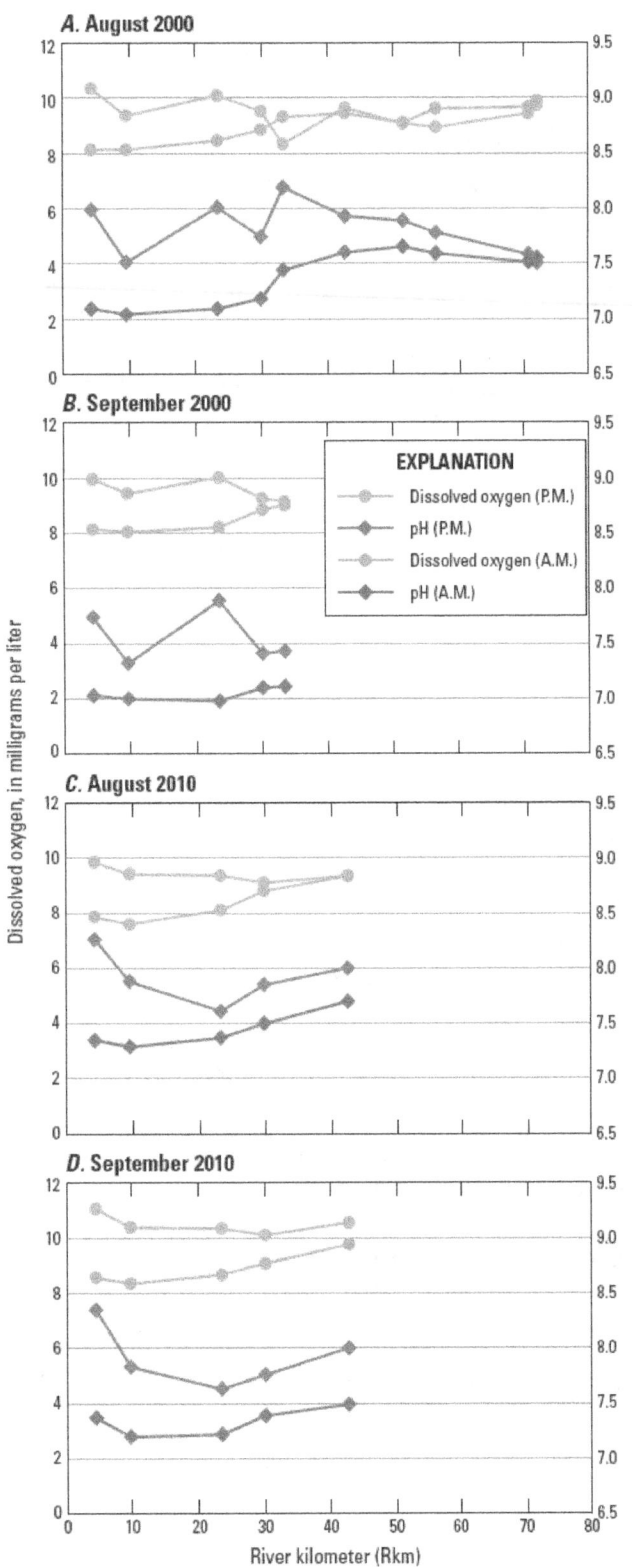

Figure 26. Concentrations of dissolved oxygen and pH levels in the Molalla River, Oregon in (*A*) August 2000, (*B*) September 2000, (*C*) August 2010, and (*D*) September 2010.

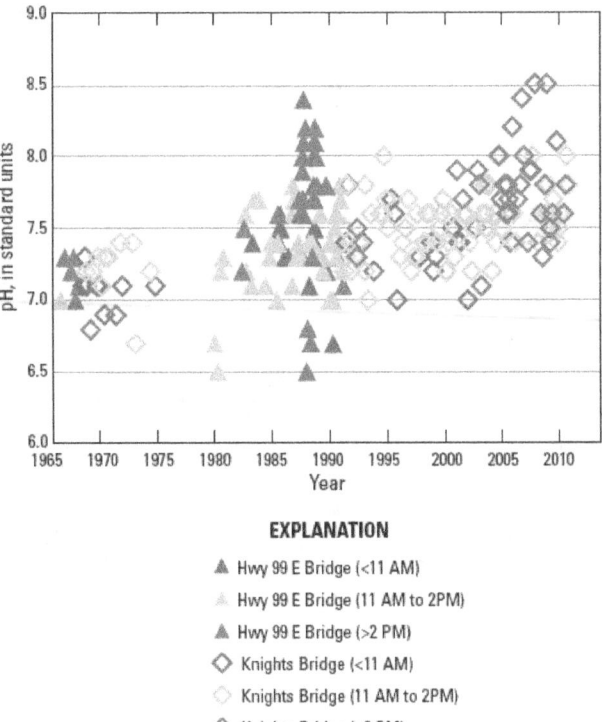

Figure 27. Long-term trend in pH values in the lower Molalla River, 1965–2010. (Data from the Oregon Department of Environmental Quality's LASAR web application.)

Nutrients

Water samples for analyses of nutrients were collected and processed using USGS protocols (U.S. Geological Survey, 2006; Wilde and others, 2004) whereby width-integrated samples were collected from beneath the water surface at multiple locations in the cross section. Samples for analyses of dissolved ammonia, dissolved nitrite-plus-nitrate (hereafter referred to as nitrate), and soluble reactive phosphorus (SRP) were filtered through 0.45-µm Acrodisk™ filters using an acid-cleaned 60 mL syringe, and total nutrient samples were acidified with 1-mL of 4.5-N sulfuric acid per each 125-mL sample. Nutrient samples were shipped on ice to the USGS National Water-Quality Laboratory in Denver, Colorado, and analyzed using USGS protocols.

Compared with many other area rivers, nutrient concentrations in the Molalla River were generally low at most sites but did exhibit a pronounced increase in both nitrogen and phosphorus at Goods and Knights Bridges, presumably a consequence of greater inputs of nutrients from anthropogenic sources in the lower basin that entered the river from tributaries, agricultural irrigation returns, or groundwater. Most of the nitrogen in the Molalla River was in the form of either dissolved organic nitrogen (upper basin sites) or

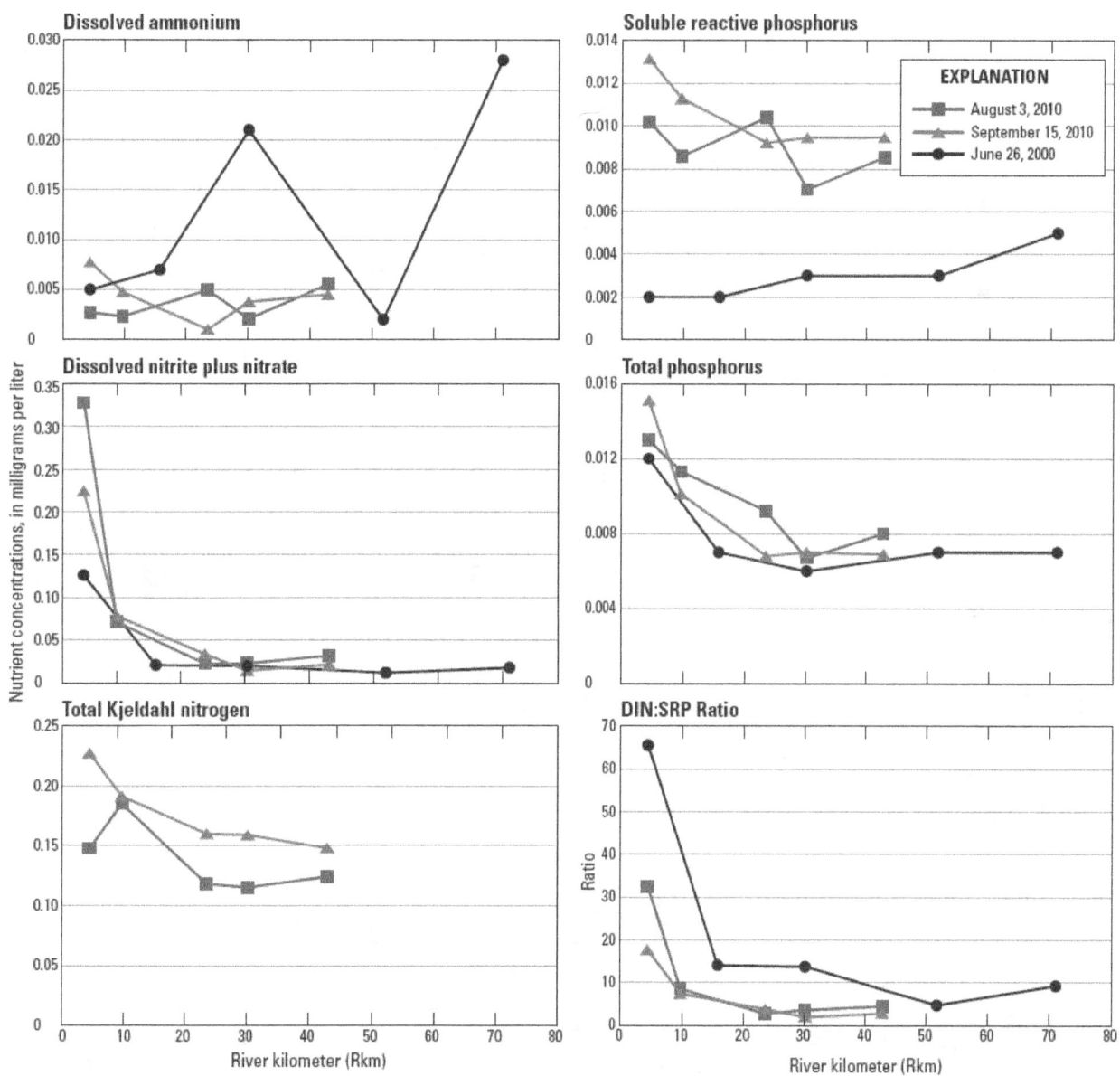

Figure 28. Longitudinal nutrient concentrations in the main-stem Molalla River, Oregon, August and September 2010 and June 2000.

dissolved nitrate (lower basin sites). Concentrations of nitrate in the Molalla River at Glen Avon Bridge (0.022–0.032 mg/L, fig. 28; table 9), although not exceptionally high, exceeded the seasonal average concentration proposed by the U.S. Environmental Protection Agency for reference streams in the Cascade Range of 0.005 mg/L (table 10). The total phosphorus (TP) concentrations at this site were slightly less than the reference concentration of about 0.010 mg/L. Near the end of the study reach at Knights Bridge, the nitrate concentrations were 0.22–0.33 mg/L (table 9), which exceeded by about 25 percent the seasonal average concentration for reference streams in the Willamette Valley of 0.15 mg/L (table 10);

TP concentrations (0.013–0.015 mg/L) were well below the reference value of 0.04 mg/L. The SRP concentrations were higher in 2010 than in 2000 (fig. 28), possibly because of lower streamflow: 2.4–3.2 m^3/s (85–113 ft^3/s) at Goods Bridge in 2010 compared to 8 m^3/s (283 ft^3/s) in 2000 (fig. 24). The relatively low phosphorus concentrations in the Molalla River reflected the lack of naturally occurring geologic phosphorus in the upper basin that is found in the upper reaches of other Cascade Range rivers influenced by High Cascades geology, such as the nearby Clackamas River, where phosphorus levels during summer can be nearly twice as high (Carpenter, 2003).

Table 9.   Nutrient concentration data for the main-stem Molalla River, Oregon, 2010.

[Nutrient concentrations in milligrams per liter. Concentrations greater than the laboratory detection limits are shown in **bold**. Nutrient data from June 2000 (U.S. Geological Survey, unpub. data) are provided for comparison. **Abbreviations:** DIN, dissolved inorganic nitrogen [nitrite+nitrate+ammonia]; SRP, soluble reactive phosphorus; TP, total phosphorus; MDL, method detection limit; –, no data]

Site	Date	River kilometer	Dissolved ammonium	Dissolved nitrite plus nitrate	Dissolved inorganic nitrogen	Dissolved organic nitrogen	Total Kjeldahl nitrogen	Total dissolved phosphorus	Soluble reactive phosphorus	Total phosphorus	DIN:SRP	SRP:TP
Molalla River upstream of Scorpion Creek	June 26, 2000	71.0	**0.028**	**0.018**	**0.046**	–	–	–	0.005	0.007	9.2	0.7
Molalla River downstream of Pine Creek	June 26, 2000	51.7	0.002	0.012	0.014	–	–	–	0.003	0.007	4.7	0.4
Molalla River at Highway 211	June 26, 2000	30.1	**0.021**	**0.020**	**0.041**	–	–	–	0.003	0.006	13.7	0.5
Molalla River upstream of Milk Creek	June 26, 2000	15.8	0.007	0.021	0.028	–	–	–	0.002	0.007	14.0	0.3
Molalla River at Goods Bridge	June 26, 2000	9.7	–	–	–	–	–	–	–	–	–	–
Molalla River downstream of Knights Bridge	June 26, 2000	4.3	0.005	**0.126**	**0.131**	–	–	–	0.002	**0.012**	65.5	0.2
Molalla River at Glen Avon Bridge	Aug. 3, 2010	42.8	0.006	**0.032**	**0.038**	0.097	**0.12**	0.005	**0.009**	**0.008**	4.4	1.1
Molalla River at Highway 211	Aug. 3, 2010	30.1	0.002	**0.023**	**0.025**	**0.102**	**0.12**	0.004	0.007	0.007	3.6	1.1
Molalla River at Highway 213	Aug. 3, 2010	23.5	0.005	**0.023**	**0.028**	0.087	**0.12**	0.003	**0.010**	**0.009**	2.7	1.1
Molalla River at Goods Bridge	Aug. 3, 2010	9.7	0.002	**0.071**	**0.073**	**0.118**	**0.19**	**0.006**	**0.009**	**0.011**	8.5	0.8
Molalla River downstream of Knights Bridge	Aug. 3, 2010	4.3	0.003	**0.327**	**0.330**	**0.139**	**0.15**	**0.007**	**0.010**	**0.013**	32.4	0.8
Molalla River at Glen Avon Bridge	Sept. 15, 2010	42.8	0.005	**0.022**	**0.026**	0.028	**0.15**	0.004	**0.009**	0.007	2.8	1.4
Molalla River at Highway 211	Sept. 15, 2010	30.1	0.004	0.014	0.018	0.005	**0.16**	0.003	**0.009**	0.007	1.9	1.4
Molalla River at Highway 213	Sept. 15, 2010	23.5	0.001	**0.034**	**0.035**	0.04	**0.16**	0.003	**0.009**	0.007	3.8	1.4
Molalla River at Goods Bridge	Sept. 15, 2010	9.7	0.005	**0.078**	**0.083**	0.044	**0.19**	**0.006**	**0.011**	0.010	7.3	1.1
Molalla River downstream of Knights Bridge	Sept. 15, 2010	4.3	0.008	**0.224**	**0.232**	0.06	**0.23**	**0.008**	**0.013**	**0.015**	17.7	0.9
Laboratory MDLs			0.02	0.016	–	0.1	0.1	0.006	0.008	0.008	–	–

Table 10.　Nutrient concentrations for reference streams in the Cascade Range and Williamette Valley, Oregon, and nutrient and algal criteria to prevent nuisance conditions in streams.

[Nutrient reference conditions suggested by the U.S. Environmental Protection Agency for streams in the Willamette Valley (Nutrient Ecoregion I, Subecoregion 3) and Cascade Range (Nutrient Ecoregion II, Subecoregion 4). Nutrient and algal criteria from Welch and others (1988). **Abbreviations:** < MRL, less than the method reporting level; mg/L, milligram per liter; mg/m², milligram per square meter]

Nutrient	Number of sites	Minimum value	Maximum value	Reference condition (seasonal average) (mg/L)
Nutrient reference conditions (Willamette Valley)				
Nitrite plus nitrate (NO_2+NO_3)	85	20	8,640	0.15
Total Kjeldahl nitrogen (TKN)	96	50	2,750	0.21
Total nitrogen (TN)	13	0	2,990	0.32
Total phosphorus (TP)	138	2	816	0.04
Nutrient reference conditions (Cascade Range)				
Nitrite plus nitrate (NO_2+NO_3)	75	< MRL	1.910	0.005
Total Kjeldahl nitrogen (TKN)	65	< MRL	0.950	0.05
Total nitrogen (TN)	27	< MRL	2.860	0.055
Total phosphorus (TP)	95	< MRL	0.243	0.009

Nutrient enrichment indicators	Criteria	Sites (dates) exceeding criteria	Maximum value
Nutrient and algal criteria to prevent nuisance conditions			
Dissolved inorganic nitrogen (NO_3+NH_4)	0.1 mg/L	Molalla River at Knights Bridge (Aug. 03 and Sept. 9, 2010)	0.327 mg/L
Soluble reactive phosphorus (SRP)	0.01–0.015 mg/L	Molalla River at Knights Bridge (Aug. 03 and Sept. 9, 2010)	0.013 mg/L
		Molalla River at Goods Bridge (Aug. 03 and Sept. 9, 2010)	0.011 mg/L
Total phosphorus (TP)	0.01–0.09 mg/L	Molalla River at Knights Bridge (Aug. 03 and Sept. 9, 2010)	0.015 mg/L
		Molalla River at Goods Bridge (Aug. 03 and Sept. 9, 2010)	0.011 mg/L
Benthic algae (periphyton) chlorophyll-*a*	100–150 mg/m²	Molalla River at Knights Bridge (Aug. 15, 2010)	113 mg/m²

Biologically available forms of dissolved inorganic nitrogen (DIN), which include both ammonium and nitrate, generally were less than 0.04 mg/L and SRP concentrations were less than about 0.01 mg/L at the two most upstream sites (table 9). This resulted in low DIN: SRP ratios, ranging from 1.9 to 4.4, which suggests that algae might be limited by nitrogen availability considering that ratios of about 7.2 indicate balanced nutrient availability, according to the "Redfield Ratio" (Redfield, 1934; Redfield, 1958; Hillebrand and Sommer, 1999). Nitrate concentrations at the downstream sites, however, increased 2–10 fold compared with upstream sites, resulting in DIN:SRP ratios up to 32 and 18 at Knights Bridge in August and September, respectively (fig. 28; table 9). These ratios suggest a possible switch from limitation by nitrogen to phosphorus limitation in the lower river, where heavy growths of filamentous green algae (*Cladophora glomerata*) were observed (see photograph 4, p. 54). It is also possible that invertebrate grazing or some other factor limits the growth of algal populations, a topic discussed later in this report.

Photograph 4.   *Cladophora glomerata*, filamentous green algae in the lower Molalla River near Canby. (Photograph taken by Jonathan Czuba, July 14, 2010).

Although water-column nitrogen is probably sufficient to saturate growth rates in the lower Molalla River, this could be an important factor governing the algal biomass and (or) species composition in the middle reaches of the river. It should be pointed out, however, that although concentrations of dissolved nutrients are often used in water-quality assessments to identify the potential for nuisance algal growth, measured concentrations may not provide an accurate assessment of stream conditions because such measures may represent only the nutrients that remain after uptake by algae. Water-column concentrations of biologically available dissolved nutrients during summer can therefore underrepresent true enrichment from nutrients when algal abundance is high because low concentrations of dissolved nutrients can result from such uptake (Mulholland and Rosemond, 1992; Dodds, 1993; Peterson and others, 2001; Carpenter, 2003). To address this issue, other indicators of eutrophication, such as benthic algal biomass, diel fluctuations in pH and DO, and algal species and diatom composition can provide a more thorough, time-integrated assessment of stream conditions.

Analysis of ODEQ's ambient monitoring data for the lower Molalla River show an apparent decline in the minimum concentrations of DIN in the past decade or so (fig. 29). Although lower DIN concentrations could be a sign of improved water quality, this trend also might be the result of greater uptake of nitrogen (primarily nitrate) by periphytic algae in the lower river, which use DIN and SRP for growth. Summertime DIN concentrations started to decline in 1997, the year after the floods of 1996 that caused widespread damage to streamside vegetation, reworked large gravel bars, and widened stream channels (figs. 16 and 17), and may have created additional habitat for benthic algae to colonize and grow over time. The observed reduction in shade in both flood plain and island bars, shown by the increase in area of bare bars from 1994 to 2000 (fig. 19) would have provided more light for algal photosynthesis in the lower river, particularly upstream of Goods Bridge. Although greater amounts of photosynthesis could have produced the observed decrease in DIN and higher pH values in recent years (fig. 27), as described above, more of the recent samples were collected after 2:00 P.M., which, given the diel cycle of algal photosynthesis, probably also contributed to these patterns.

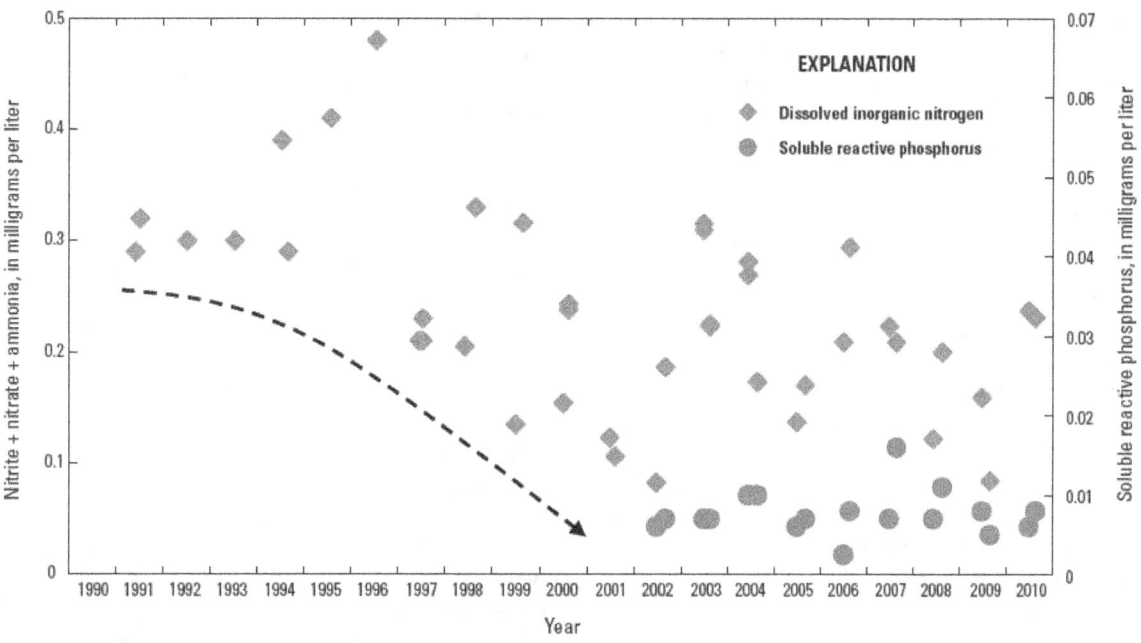

Figure 29. Summertime dissolved inorganic nitrogen and phosphorus concentrations in the Molalla River, Oregon, 1990–2010.

Benthic Community Conditions—Algae and Macroinvertebrates

Benthic community conditions were assessed in the Molalla River in late July and mid-August in 2010. Benthic algae samples were collected from rock and cobble substrates in riffle areas and analyzed for biomass (chlorophyll-*a*, or Chl. *a*) and species composition. To randomize rock selection, flat metal washers with short lengths of orange flagging were hand-tossed from shore to 10 locations within the sampling zone, and the rock or cobble immediately upstream of the washer was carefully removed and brought to the bank, where algal material was collected using the top-rock/cylinder scrape method (Moulton and others, 2002). The top of each rock was covered with a plastic PVC cylinder to form a scribe having outside diameters ranging from 4 to 10.4 cm. The largest scribe possible was used on each rock to maximize the surface area sampled, and algal material outside the scribe was removed with a knife and/or plastic bristle brush and discarded. The circular patch of algae remaining on each rock was then scraped and rinsed with native water into a plastic basin, compositing approximately 1–2 L of algal material per sample. The algae and rinse-water slurry was transferred into 1-L Nalgene® bottles and placed on ice until further processing. Periphyton biomass samples were processed off-site by homogenizing in an electric blender and transferring into a churn splitter. Subsamples of 5–10 mL for analysis of Chl. *a* were removed from the churn using a large-orifice pipettor and transferred onto 47-mm glass-fiber (GF/F) filters under vacuum pressure using a plastic filtration apparatus. Chl. *a* was analyzed at the Oregon Water

Science Center laboratory in Portland, Oregon using a Turner Designs fluorometer with acid correction for the presence of phaeopigments.

Samples for algae species identification and enumeration were obtained from the same churn splitter as the biomass samples, preserving 95 mL of the algal slurry material with 5 mL of full strength buffered formalin at 5 percent final preservative concentration. Samples were analyzed using methods described in Aquatic Analysts (2007). Cell density enumerations and biovolume measurements of a minimum of 100 live algal units were made under a Zeiss standard microscope equipped with phase contrast up to 1,000X magnification. Only viable cells, verified by intact chloroplasts, were counted. Algae were enumerated along a linear transect, measured accurately to 0.1 mm with a stage micrometer. Cell densities were calculated from the area observed (transect length × diameter of field of view), and taking into account the effective filter area and the volume of sample filtered used in sample preparation. Average biovolume estimates of each species were obtained from measurements on each taxon encountered in each sample.

Additional algal samples were removed from the same churn or collected separately at a site for supplemental qualitative microscopic evaluations of unpreserved material. Samples were examined with a Leica DM1000 light microscope equipped with phase contrast using a range of magnifications up to 1000X. These observations complemented the algal identifications provided by Aquatic Analysts by identifying to genus or species many of the larger algal forms, or "macro algae" that can be missed in the cell counts.

Benthic algal biomass values in the Molalla River were moderately high in July, ranging from 30 to 55 mg chlorophyll-*a*/m² and did not vary much longitudinally (fig. 30). Much of the river channel was covered with a glistening layer of diatoms and other algae (see photographs 5 and 6, p. 57). In August, biomass levels were higher at the Goods and Knights Bridges sites than at the three upstream sites. A large population of *Cladophora glomerata*, a high-biomass forming type of filamentous green algae, developed in the lower river in July, producing a biomass of 113 mg chlorophyll-*a*/m² downstream of Knights Bridge in August, which exceeds the 100 mg/m² threshold commonly used to indicate nuisance conditions (Welch and others, 1988). Although *Cladophora* was present in the riffle sampled at the Knights Bridge site, the abundance was much higher in the deeper, slower flow run habitats in the reach downstream (see photograph 4, p. 54). *Cladophora* also produces nuisance conditions in other Cascade Range streams, including the nearby Clackamas River (Carpenter, 2003), the North Umpqua River in southwestern Oregon (Anderson and Carpenter, 1998), and many other streams. As described further in the Evaluation of Quality Assurance Data section, because of the limited cell counts and issues related to either subsampling during sample preparation or magnification, some types of macroalgae, including filamentous green algae, were not detected during the cell counts but were observed in the river and noted during sampling (table 11). *Cladophora*, for example, was not identified in the cell counts from Knights Bridge, but was abundant at this and other downstream locations (see photograph 4, p. 54), and was observed in the supplemental qualitative microscopic evaluations of samples collected at other sites, including those upstream of Glen Avon Bridge.

Benthic algal assemblages in the Molalla River included small, fast growing diatoms and very large stalked diatoms, filamentous green and blue-greens, and a few planktonic forms of green and blue-green algae (appendixes B and C). Just 33 algal taxa, mostly pennate diatoms, were identified in microscopic cell counts. Compared with other streams and rivers in the Pacific Northwest (Cuffney and others, 1997; Anderson and Carpenter, 1998; Carpenter, 2003; Waite and others, 2008), species richness in the Molalla River was rather low, averaging just 11–17 taxa per sample. This is partly due to the limited number of cells counted—100 "algal units" per sample—but also may reflect the similar condition of habitat sampled (riffle cobbles), particularly for sites between the Glen Avon and Goods Bridges.

An algal autecological indicator species analysis was conducted on the diatom assemblages from the Molalla River to gage the degree of nutrient and organic enrichment using species classifications, preferences, and tolerances published in Porter (2008). Table 12 lists all 33 algal taxa identified in cell counts and their respective autecological classifications for nutrients, dissolved oxygen, pH, salinity, and organic pollution. Table 13 lists the percentages of each indicator group (guild) that were tallied for each sample, considering only those taxa classified for a particular trait. For example, the percentage of eutrophic taxa was computed as the proportion of diatom taxa classified as eutrophic divided by the total percentage of diatoms classified for all trophic categories. Indicator percentages were computed on the basis of the cell density and algal biovolume for the July and August samplings separately, and averages for each time period are shown at the bottom of table 13. The average percent abundances for the combined percent density and biovolume of each guild for July and August are shown in figure 31.

The two most abundant guilds were pollution sensitive diatoms intolerant of excessive organic enrichment (Bahls, 1993) and diatoms requiring high concentrations of dissolved oxygen (table 12). These guilds comprised 50 percent or more of the relative abundance in both July and August samples (fig. 31). Five diatoms (*Achnanthes minutissima* [*Achnanthidium minutissimum*], *Achnanthes linearis*, *Gomphonema subclavatum*, *G. angustatum*, and *Cymbella minuta*) occurred in all samples (appendix B and appendix C), making up, on average, 77 percent of the cell density and 34 percent of the biovolume per sample. All five taxa are sensitive to organic pollution, which is consistent with the relatively low organic content of the Molalla River during summer. Sedimentation of riffles does not appear to be an issue as motile diatoms—those that often thrive in habitats affected by sedimentation—constituted less than 2 percent of taxa in all samples (not shown), consistent with the relatively "clean" riffles sampled that had low substrate embeddedness, generally less than 5 percent, indicating only minor infilling of fine sediments among the riffle cobbles. More fine sediments in the lowermost site at Knights Bridge resulted in somewhat higher substrate embeddedness of about 10–25 percent.

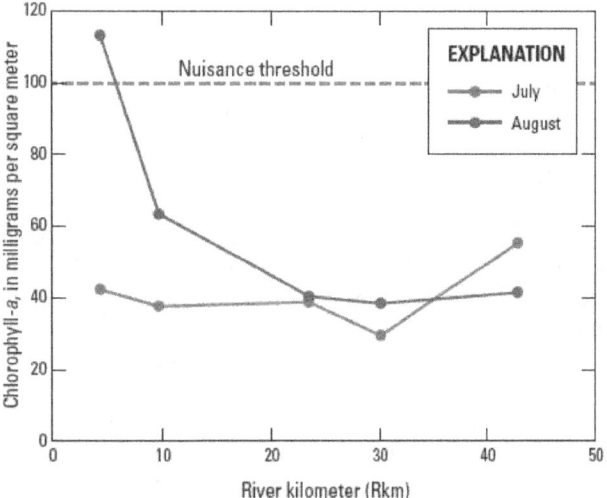

Figure 30.   Longitudinal pattern in periphyton biomass (chlorophyll-*a*) in the main-stem Molalla River, Oregon, July and August 2010.

Photographs 5 and 6.   Diatom biofilm on a rock collected from the Molalla River upstream of Glen Avon Bridge, and resulting slurry of algal material. (Photographs taken by Kurt Carpenter, July 30, 2010).

Table 11. Periphyton biomass and qualitative descriptions of benthic communities in the main-stem Molalla River, Oregon, July–August 2010.

[Algal biomass estimated from benthic chlorophyll-*a*. **Abbreviation:** mg/m², in milligram per square meter]

Date	Site	River kilometer	Chlorophyll-*a* (mg/m²)	Qualitative description of benthic periphyton	Qualitative description of benthic invertebrates
July 30, 2010	Molalla River upstream of Glen Avon Bridge	42.8	55	Moderately thick diatom biolfilm with some tufts of filamentous green algae.	Numerous caddisflies (*Dicosmoecus* and Glossosomidae), mayflies (Heptageniidae), and stoneflies (*Calaneria*), and numerous midges (Chironomidae).
July 30, 2010	Molalla River upstream of Highway 211 Bridge	30.1	30	Thick diatom biofilm.	Numerous caddisflies (*Dicosmoecus* and Glossosomidae), mayflies (Heptageniidae), and stoneflies (*Calaneria*), and numerous midges (Chironomidae).
July 30, 2010	Molalla River upstream of Highway 213 Bridge	23.5	39	Moderately thick diatom biofilms and filamentous blue green mats (*Oscillatoria sp.*) starting to flake off.	Numerous caddisflies (*Dicosmoecus* and Glossosomidae), mayflies (Heptageniidae), and stoneflies (*Calaneria*), and numerous midges (Chironomidae).
July 30, 2010	Molalla River upstream of Goods Bridge	9.7	38	Moderately thick diatom biolfilm.	Snails and midges (Chironomidae) abundant, with fewer mayflies (Heptageniidae), stoneflies (*Calaneria*), caddisflies (Glossosomidae).
July 30, 2010	Molalla River downstream from Knights Bridge	4.3	42	Large growths of *Cladophora glomerata*.	Snails and midges (Chironomidae) abundant, with fewer caddisflies (Glossosomidae) and stoneflies.
Aug. 15, 2010	Molalla River upstream of Glen Avon Bridge	42.8	41	Thin diatom biofilm with filamentous blue-green algae, clumps of stalked diatoms, and lesser filamentous green algae.	Fewer benthic invertebrates; a few stoneflies and caddisflies (pupating).
Aug. 15, 2010	Molalla River upstream of Highway 211 Bridge	30.1	38	Moderately thick diatom biofilm with lesser filamentous blue-green and filamentous green algae.	Fewer benthic invertebrates; a few stoneflies, mayflies (Heptageniidae), and caddisflies (pupating).
Aug. 15, 2010	Molalla River upstream of Highway 213 Bridge	23.5	40	Moderately thick diatom biofilm with lesser filamentous blue-green and filamentous green algae.	Fewer benthic invertebrates; a few stoneflies, mayflies (Heptageniidae), and pupating caddisflies (Glossosomid and Dicosmoecus).
Aug. 15, 2010	Molalla River upstream of Goods Bridge	9.7	63	Moderately thick diatom biofilm with clumps of stalked diatoms; extensive growth of filamentous green algae along right bank upstream of the bridge.	Snails abundant with a few stoneflies and caddisflies (Glossosomidae).
Aug. 15, 2010	Molalla River downstream from Knights Bridge	4.3	113	Moderately thick diatom biofilm with some decaying filamentous green algae.	Snails abundant.

Table 12. Autecological preferences and indicator qualities of algal taxa identified in the Molalla River, Oregon, July and August 2010.

[Taxa listed in descending order according to the total biovolume for each taxa considering all samples. Autecology from Porter, 2008. Bahls' pollution tolerance from Bahls (1993). **Abbreviations**: sp., species; NF, nitrogen fixer; E, eutrophic; M, mesotrophic; ME, meso-eutrophic; O, oligotrophic; AH, always high (~100 percent saturation); FH, fairly high (>75 percent saturation); MOD, moderate (>50 percent saturation); L, LOW (>30 percent saturation); ALK-P, alkaliphilic (pH >7); ALK-B, alkalibiontic (pH >7); INDIFF, indifferent; CIR, circumneutral (pH ~7); INDIFF, indifferent to pH; FB, fresh-brackish; F, freshwater; BF, brackish-fresh; S, sensitive; LT, less tolerant; VT, very tolerant; –, unknown; ~, approximate; >, greater than]

| Algal taxa | Algal division | Number of samples | Autelogical preference or indicator quality | | | | | |
			Nitrogen fixer	Trophic (nutrient indicator)	Dissolved oxygen	pH	Salinity	Bahls' pollution tolerance
Cymbella affinis	Diatom	9	No	E	AH	ALK-P	FB	S
Synedra ulna	Diatom	8	No	–	MOD	ALK-P	FB	LT
Achnanthes linearis	Diatom	10	No	–	–	CIR	–	S
Achnanthes minutissima	Diatom	10	No	–	AH	INDIFF	FB	S
Gomphonema subclavatum	Diatom	10	No	OM	AH	CIR	FB	LT
Gomphonema angustatum	Diatom	10	No	O	AH	ALK-P	FB	LT
Gomphoneis herculeana	Diatom	2	No	–	–	–	–	S
Cymbella minuta	Diatom	10	No	–	–	CIR	FB	LT
Cymbella cistula	Diatom	3	No	E	FH	ALK-P	FB	S
Cocconeis placentula	Diatom	8	No	E	MOD	ALK-P	FB	S
Fragilaria vaucheriae	Diatom	7	No	E	MOD	ALK-P	FB	LT
Cymbella tumida	Diatom	2	No	ME	AH	ALK-P	FB	S
Gomphonema ventricosum	Diatom	4	No	O	AH	–	F	–
Aphanizomenon flos-aquae	Blue-green	2	Yes	E	–	ALK-B	–	–
Synedra rumpens	Diatom	6	No	–	–	INDIFF	–	LT
Scenedesmus quadricauda	Green	3	No	E	–	ALK-P	–	–
Gomphonema tenellum	Diatom	5	No	–	–	–	–	S
Cymbella sp.	Diatom	3	No	–	–	–	–	–
Nitzschia frustulum	Diatom	4	No	E	MOD	ALK-P	BF	LT
Cymbella sinuata	Diatom	4	No	M	AH	CIR	FB	S
Gomphonema clevei	Diatom	4	No	–	–	–	–	S
Synedra tenera	Diatom	1	No	OM	AH	ALK-P	F	–
Navicula cryptocephala veneta	Diatom	5	No	E	L	ALK-P	BF	LT
Navicula graciloides	Diatom	1	No	E	AH	ALK-B	FB	–
Nitzschia dissipata	Diatom	1	No	ME	FH	ALK-P	FB	S
Nitzschia palea	Diatom	1	No	E	L	CIR	FB	VT
Rhoicosphenia curvata	Diatom	2	No	E	FH	ALK-P	FB	S
Gomphonema olivaceum	Diatom	1	No	E	FH	ALK-B	FB	S
Nitzschia microcephala	Diatom	1	No	E	MOD	ALK-P	FB	VT
Achnanthes hauckiana	Diatom	1	No	–	–	–	–	LT
Ankistrodesmus falcatus	Green	1	No	E	–	ALK-P	–	–
Navicula cryptocephala	Diatom	1	No	–	MOD	ALK-P	FB	S
Navicula decussis	Diatom	1	No	ME	–	ALK-P	FB	S

Table 13. Diatom indicator species (guilds) in the Molalla River, Oregon, July and August 2010.

[Percent relative abundance for each periphyton guild. Guild information from Porter (2008). Replicate samples from Goods Bridge in August were averaged. Bahls' pollution tolerance from Bahls (1993)]

Date	Site	Abundance expression (percent)	Nutrients		Oxygen and pH			Lange-Bertalot pollution tolerance		Bahls' pollution tolerance	
			(Low) Oligotrophic	(High) Eutrophic	High dissolved oxygen	Low-moderate dissolved oxygen	High pH	Less tolerant	Tolerant-very tolerant	Pollution sensitive	Pollution tolerant
July 30, 2010	Molalla River at Glen Avon Bridge	Biovolume	14	68	75	25	61	80	20	70	0.0
	Molalla River at Highway 211		9.1	78	76	24	89	77	23	70	0.0
	Molalla River at Highway 213		7.0	89	39	51	90	49	51	45	0.0
	Molalla River at Goods Bridge		8.4	88	80	0.3	74	99	0.3	88	0.0
	Molalla River downstream of Knights Bridge		7.4	64	43	56	62	49	50	42	0.0
	Molalla River at Glen Avon Bridge	Density	41	38	94	6.3	18	98	2.2	85	0.0
	Molalla River at Highway 211		37	51	90	10	42	93	7.8	76	0.0
	Molalla River at Highway 213		31	62	78	21	38	78	20	68	0.0
	Molalla River at Goods Bridge		44	44	98	1.0	20	98	1.0	85	0.0
	Molalla River downstream of Knights Bridge		19	57	68	29	27	82	17	69	0.0
Aug. 15, 2010	Molalla River at Glen Avon Bridge	Biovolume	14	62	76	24	59	94	6.3	60	0.3
	Molalla River at Highway 211		7.3	77	45	40	69	59	37	48	0.0
	Molalla River at Highway 213		11	57	79	20	63	75	20	68	0.0
	Molalla River at Goods Bridge		26	39	77	23	45	74	25	59	0.2
	Molalla R. downstream of Knights Bridge		13	49	89	11	40	83	17	65	0.5
	Molalla River at Glen Avon Bridge	Density	27	55	76	23	37	91	8.6	65	0.8
	Molalla River at Highway 211		28	56	80	19	27	82	14	67	0.0
	Molalla River at Highway 213		22	59	82	16	21	76	13	74	0.0
	Molalla River at Goods Bridge		47	31	92	7.0	18	93	5.6	79	0.7
	Molalla River downstream of Knights Bridge		41	21	94	6.0	14	92	7.7	77	0.8
Averages	**July**	**Biovolume**	**9**	**77**	**63**	**31**	**75**	**71**	**29**	**63**	**0.0**
	August		**14**	**57**	**73**	**24**	**55**	**77**	**21**	**60**	**0.2**
	July	**Density**	**35**	**50**	**85**	**14**	**29**	**90**	**10**	**76**	**0.0**
	August		**33**	**44**	**85**	**14**	**23**	**87**	**10**	**72**	**0.5**
Overall averages			23	57	76	21	46	81	17	68	0.2

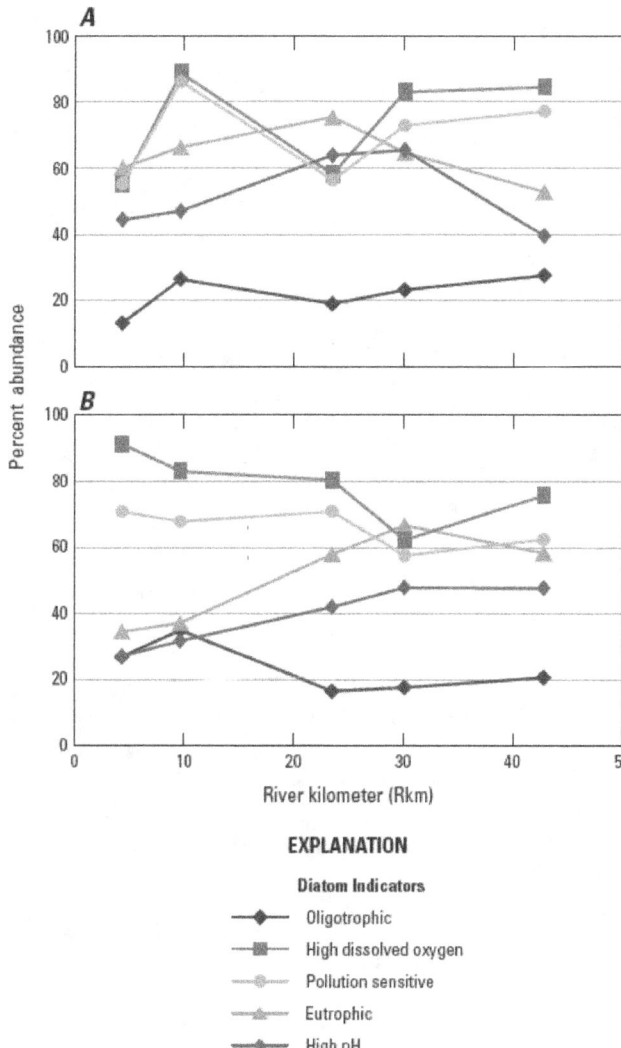

high dissolved oxygen indicators occurred in this reach in July (fig. 31), largely due to increases in two *Cymbella* species, *C. affinis* and *C. cistula* (appendixes B and C). The "improvement" in algal indicators at Goods Bridge is somewhat unexpected, given that water temperature, specific conductance, and nitrate concentrations were higher, and dissolved oxygen concentrations were lower at Goods Bridge than at the Highway 213 site, although the differences were small (less than 10 percent), except for nitrate, which was 2–3 times higher at Goods Bridge. Such discrepancies between measured and inferred water quality from algal indicators can sometimes occur because algae integrate conditions over time (Carrick and others, 1988; Lowe and Pan, 1996), whereas water samples characterize conditions only at the time of sampling.

Compared with other guilds, the abundance of oligotrophic diatoms, or taxa indicative of low nutrient conditions, was less than 15 percent of the average biovolume and approximately 35 percent of the average density in July and August (table 13). Although higher abundances of oligotrophic taxa were observed at individual sites— the highest abundance was at Goods Bridge from two *Gomphonema* species (*G. angustatum* and *G. ventricosum*)— they never comprised more than half of the relative density or biovolume in a sample (table 12). Oligotrophic taxa are typical of low-nutrient "reference" streams, so their relatively low abundance in the Molalla River suggests some degree of nutrient enrichment, despite the apparently low nutrient levels measured.

In contrast to many other Cascade Range rivers such as the Clackamas or North Umpqua Rivers (Anderson and Carpenter, 1998; Carpenter, 2003), the abundance of nitrogen-fixing taxa were notably low in the Molalla River. As the sole exception, *Aphanizomenon flos-aquae*, a colonial blue-green algae, was identified in samples from the Highways 211 and 213 sites in July. *Aphanizomenon* is typically planktonic and occurs in lakes, reservoirs, and other ponded water bodies, where it sometimes forms blooms, so its occurrence in the Molalla River periphyton is unusual. It is possible that the *Aphanizomenon* originated from a bloom in a stagnant side channel or alcove; drainage from off-channel ponds is another possible source.

The absence of *benthic* nitrogen fixing algae combined with the relatively low concentrations of SRP and detectable nitrogen in the form of nitrate suggest that algae in the lower river are probably not limited by nitrogen. The occurrence of filamentous green algae, a type that requires high amounts of nitrogen, also supports this hypothesis.

Other types of macroalgae, including the large stalked diatoms *Gomphoneis herculeana* and *Cymbella cistula*, and the filamentous blue-green algae *Oscillatoria* sp. were observed in the supplemental qualitative microscopic evaluations. The occurrence of *G. herculeana*, *C. cistula*, and other stalked diatoms in the Molalla River is noteworthy because they are becoming increasingly notorious for

EXPLANATION

Diatom Indicators

♦ Oligotrophic
■ High dissolved oxygen
● Pollution sensitive
▲ Eutrophic
♦ High pH

Figure 31.   Longitudinal pattern in diatom indicators of water-quality conditions in the Molalla River, Oregon, in (*A*) July and (*B*) August 2010.

Other types of diatoms in the Molalla River are, however, indicative of waters having high levels of nutrients (=eutrophic, meaning "well fed") and alkaline pH (greater than 7 standard units). Many species of algae, including several types of filamentous green algae such as *Cladophora*, *Stigeoclonium*, and *Zygnema*, and many diatoms require high nutrient levels. Eutrophic diatoms constituted between 77 and 57 percent of the average algal biovolume and between 50 and 44 percent of the average cell density, considering all samples in July and August, respectively (table 13). High-pH-indicating diatoms constituted about 25 to 65 percent of the relative abundance, and exhibited a modest downstream decline in value along with the percent eutrophic diatoms between the Highway 213 and Goods Bridges (fig. 31). Increases in oligotrophic, pollution sensitive, and fouling streams with high biomass and producing nuisance conditions (Spaulding and Elwell, 2007). In New Zealand, and in Montana and other northwestern states, a similar type of stalked diatom (*Didymosphenia geminata*, so-called "rock snot") is blamed for decimating fish populations in blue-ribbon trout streams by smothering benthic invertebrate populations with a felt-like blanket of stalk material. Although not yet identified in the Molalla River or the Clackamas River to the north, *Didymosphenia* has been identified in the North Santiam River to the south of the Molalla River (USGS, unpublished data). The stalks of these diatoms are made of polysaccharide mucilage (Wustman and others, 1997) that can become thick (as much as 20 cm), covering whole areas with white-to-brown or golden colonies. Although not overly abundant in the Molalla River, several types of stalked diatoms were identified in cell counts at all sites in both July and August samples (appendixes B and C). Stalked diatoms have the potential to cause significant impacts to stream ecosystems by altering habitat, reducing food resource quality for benthic macroinvertebrates, and smothering fish spawning redds. Because these types of algae contain large masses of mucilage, they can survive partial desiccation, so management strategies to control the spread have included, for example, educational efforts aimed at anglers and other river users to encourage careful washing of wader boots or avoiding the use of felt-soled waders that retain moisture.

Although not quantified during this study, benthic macroinvertebrates were qualitatively assessed during a July 2009 float trip and again during the two algal samplings in 2010 (table 11). A previous survey of benthic macroinvertebrates by Cole (2002) found downstream declines in richness and abundance of sensitive species in the Molalla River, with lower diversity and more tolerant species in the lower basin. In 2009, a diverse benthic assemblage consisting of mayflies, stoneflies, caddisflies, and other taxa were found in riffles in the reach between the Highway 211 and Highway 213 bridges, and in 2010, benthic macroinvertebrates were abundant at all sites, with a trend toward fewer mayflies, stoneflies, and caddisflies, and more snails and Chironomid midges at the downstream sites. Very large abundances of the large stone-cased caddis fly larvae *Dicosmoecus gilvipes* were observed on the river bottom in the reach between the Glen Avon and Highway 213 bridges during mid-summer float trips in 2010 (see photographs 7–9, p. 62). *Dicosmoecus* adults have one brood per year (univoltine), and early instars that overwinter have lighter cases that make them more vulnerable to scouring high flows compared with the larger adults. If they do survive winter, larvae rebuild cases made of small stones in the summer that make them virtually invulnerable to predators, but the large size and heavy weight of their rock casings also makes it difficult for *Dicosmoecus* to find shelter in stream bed interstices during streambed-moving high flow events (Wooton and others, 1996 and references cited within).

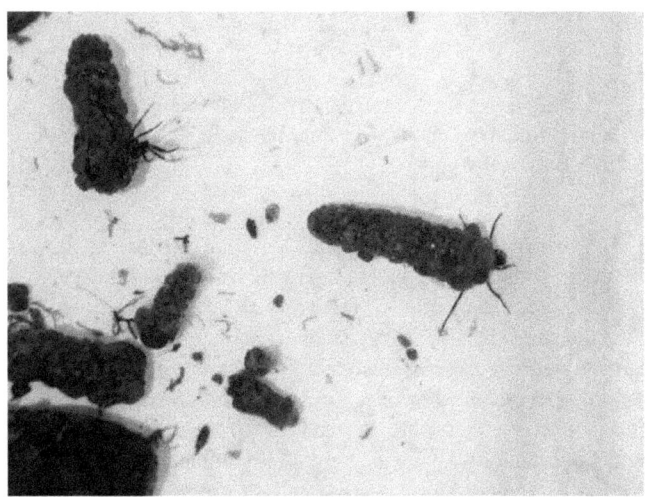

Photographs 7–9. *Dicosmoecus gilvipes* in the Molalla River. (Photographs taken by Kurt Carpenter, July 2009 and July 2010.)

The moderate abundance and patchy growth patterns of periphyton, combined with highly abundant invertebrate grazers, demonstrated the high degree of secondary production in Molalla River riffles. *Dicosmoecus* and other Limnephilid caddisflies and several types of mayflies are active grazers of benthic algae, and their high abundance in the Molalla River suggested that the accumulation of periphyton biomass may be regulated by these herbivores during summer, as was previously found in experimental stream channels (Walton and others, 1995), natural streams in the Pacific Northwest (Lamberti and others, 1992), and northern California (Wooton and others, 1996).

Synthesis of Water-Quality and Benthic Community Analyses

Overall, the water quality and algal conditions, combined with a qualitative assessment of the benthic invertebrate populations, showed that the Molalla River, although exceptionally warm during summer, is a moderately nutrient enriched and productive stream capable of providing abundant food resources for salmonids. Despite the high amounts of primary productivity, based on the diatom species present, the river is not organically enriched. At all but the most downstream site, algal biomass levels were only moderately high, which could have resulted from high densities of grazing benthic macroinvertebrates, relatively low phosphorus levels, or some other factor.

Several factors contribute to the productive nature of the Molalla River. Nutrient levels in the river were generally moderate at all but the most downstream sites, and concentrations of biologically-available dissolved nitrate and phosphate remained above detectable levels despite the prolific growths of periphyton, indicating that the nutrient supply or regeneration is keeping up with demand by stream algae. Based on the great areal extent of periphytic growth in the river, it is expected, although not measured, that demand for dissolved nutrients would be substantial. Light is also readily available during summer when the absence of rainfall and runoff result in very clear water with low turbidity (less than 2 Formazin Nephelometric Units, or FNUs; table 8). The low flows in summer create expanses of shallow riffles that provide suitable habitat for diatoms and filamentous green algae to flourish. These algae feed a diverse assemblage of benthic macroinvertebrates and provide a solid base to the food web that is critical for fish. Although riparian vegetation is substantial in some sections, the riverbed receives direct sunlight in many areas, supporting photosynthesis and contributing to downstream warming.

Even though inclement weather and slightly higher than normal streamflows delayed the 2010 growing season, algal growths and luxuriant bio-films of diatoms were observed throughout much of the river wherever the channel was open to sunshine and water depths were shallow enough to permit light for photosynthesis. A more favorable growing season might have produced higher biomass, and larger daily swings in dissolved oxygen and pH than those recorded in 2010. Nonetheless, evidence of active photosynthesis by algae, including supersaturated concentrations of dissolved oxygen and diel fluctuations in dissolved oxygen and pH were observed throughout the study reach, even at the most upstream site upstream of Glen Avon Bridge. Much larger diel swings in dissolved oxygen levels and pH were observed downstream at Knights Bridge, where the nuisance threshold of 100 mg chlorophyll-a/m^2 was exceeded in spotty but locally heavy growths of filamentous green algae (*Cladophora glomerata*). In previous years, decaying mats of *Cladophora* have clogged the drinking-water intake for the City of Canby water treatment plant (Brian Hutchins, City of Canby, oral commun., 2011). The annual cycles and severity of growth of *Cladophora* in streams and rivers is commonly highly variable from year to year (Powers and others, 2008), depending, in part, on the degree of grazing by benthic macroinvertebrates or severity of winter floods (Wooton and others, 1996).

Geomorphic and Water-Quality Factors Affecting Algae and River Food Webs

Algae in streams are affected by light availability, nutrient supply, physical habitat conditions, and grazing by herbivorous macroinvertebrates and fish, among other factors reviewed in Stevenson and others (1996). Streamflow, channel gradient, sediment supply, and other factors dictated by the river's geomorphic framework affect the amount and quality of shallow riffle habitat suitable for periphyton to develop in the Molalla River, which in turn may influence the abundance and make up of grazer and fish populations at higher trophic levels.

Through its influence on stream channels and aquatic life in streams, flooding and associated changes in the streambed can have profound effects on the river and its ecology. Peak-flow events, particularly ones that result in mobilization of stream bed material, can alter benthic communities and riverine food webs by suppressing or releasing algal populations through physical removal mechanisms (scour by sediments) and by affecting interactions among organisms occupying multiple trophic levels including primary producers, invertebrate and fish grazers, and top level predators (Powers and others, 2008). These influences can ultimately affect algal biomass levels, which are important because some of the water-quality impacts, such as high pH or low levels of dissolved oxygen, are worse when algal biomass is high. And, as an important food resource, algal abundance can govern the success of organisms positioned higher on the food chain that in the Molalla River culminates in fish.

Floods can have direct effects on channel location and form, and channel avulsions in the Molalla River have created new channels through the flood plain, resulting in dry abandoned channels in some reaches, which have unknown impacts on fish-spawning areas. Also, the channel migration and widening that has occurred in some areas produces a less shaded river, which can allow more sunlight to penetrate, promote higher water temperatures and bacterial growth, and lower dissolved oxygen concentrations in the river. Additional sunlight also favors the development of periphyton, including nuisance types of filamentous green algae such as *Cladophora*. Although determining the cause and effect relations among the various potential factors that affect algae was beyond the scope of this study, the multivariate analyses described below did identify some combinations of geomorphic and water-quality variables that could explain a significant amount of variability in the algal assemblages.

Multivariate Analyses of Diatom Assemblages

To supplement the algal autecological indicator species analyses, non-parametric multivariate statistical techniques were performed on the diatom species composition data (cell density and algal biovolume) using PRIMER-E, version 6 (Clarke and Gorley, 2006). Such analyses help simplify the inherent complexity of the algal species data and resulted in a better understanding of patterns in the diatom assemblages among the sites or samples. These multivariate techniques allowed for testing of water quality and geomorphic variables to identify factors that may influence the diatom assemblage structure.

After removing the three non-diatom taxa from the dataset, diatom-only data were square-root transformed and Bray-Curtis similarity matrices were generated to produce a non-metric multidimensional scaling (MDS) ordination plot (fig. 32). The ordination algorithm works to position samples (points on the plot) having greater similarity closer to each other, and samples that are more dissimilar are plotted farther apart through an iterative process until the most parsimonious result is achieved. Using this method, complex multivariate data can be simplified into plots containing just two dimensionless axes.

Initially, analyses were carried out for density and biovolume separately, which produced inconsistent ordinations for the four possible combinations of cell density/biovolume for the July and August data-collection periods. In one case, the sites were arranged in a downstream pattern indicative of subtle change from one site to the next, whereas other ordinations either grouped sites inconsistently or not at all. These issues result from the fact that the algal data matrix is highly skewed by a few taxa that occur in either high density (small fast growing benthic diatoms *Achnanthes minutissima* and *Achnanthes linearis;* appendix B) or those with very large cell biovolumes of 5,000–6,000 µm³ (stalked diatoms *Gomphoneis herculeana, Cymbella cistula,* and *Cymbella tumida;* appendix C) that emphasizes their abundance in

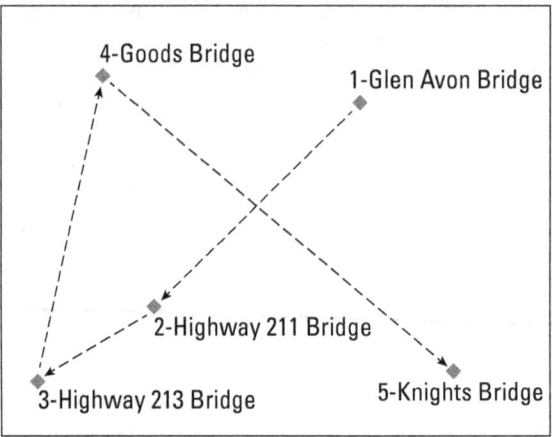

Figure 32.   Multidimensional scaling (MDS) ordination plot of algal samples from the Molalla River, Oregon, for the combined July and August 2010 samples.

analyses based on biovolume. Differences in the various ordinations (not shown) reflect that analyses based on density emphasize small taxa whereas biovolume data highlight the larger taxa. The final analysis was based on a hybrid of density and biovolume, developed by averaging the relative cell density and relative algal biovolume into one expression for the analysis. This allows for simultaneous influence by the most abundant diatoms based on size and numbers. In addition, because the most abundant taxa were found at all five sites, there was a high degree of similarity among samples from the five main-stem sites in the Molalla River, ranging from 47 to 76 percent during each sampling. Considering that percent similarities of 60–70 percent are on par with what might be expected in replicate samples, these high percent similarities reveal, along with the low species richness, that algal assemblages did not vary considerably at the sites sampled.

The resulting algal species ordination shows a trajectory in samples from Glen Avon Bridge downstream to the Highway 211 site and on to the Highway 213 site, then the trajectory changes direction whereby the Goods Bridge site plots closer to Glen Avon Bridge, whereas the Knights Bridge site is positioned out of the loop in the far corner (fig. 32). When considered in light of the results from the algal autecological analyses, the pattern in the ordination indicates that assemblages at the Goods Bridge site are more similar to those at the Glen Avon site. Higher proportions of oligotrophic, pollution-sensitive diatoms that require high levels of dissolved oxygen and lower abundances of eutrophic and high-pH-indicator diatoms were found at the Goods Bridge site (fig. 31). Actual measurements of water quality at Goods Bridge in 2010 show dissolved oxygen concentrations near saturation for much of the day, with daily minima of 85–90 percent saturation, relatively low conductance (65– 70 µS/cm), low concentrations of SRP (approximately 0.01 mg/L), and moderate concentrations of nitrate

(0.07–0.08 mg/L), but as mentioned previously, nutrient and dissolved oxygen levels measured in August and September actually indicated a small decline in quality from Highway 213 to Goods Bridge.

Associations between the algal species composition data matrix and the environmental data matrix (geomorphic and water quality variables: tables 8, 9, and 14), and select habitat data from Cole (2004) were examined using the BEST procedure in PRIMER-E. The analysis identifies variables and combinations of variables that best describe the variation in algal assemblages among sites. For this analysis, binary environmental variables including the presence of local gravel bars and bedrock were assigned a value of "1" for presence or "0" for absence, and all environmental data were log transformed and standardized, or "normalized" in PRIMER-E, prior to utilization in the BEST analysis. Because the BEST analyses were performed on just a few samples (n=5 for the average of July and August samplings), these results represent only a cursory view of the potential interactions between algal assemblages and their environment. Other factors, such as benthic invertebrate grazing, light availability, or inputs of nutrient-rich groundwater, which can be important in shaping algal assemblages (Lyford and Gregory, 1995; Stevenson and others, 1996), were not evaluated during this study or included in this analysis.

The multivariate analyses linking the periphyton species composition with a variety of environmental variables found that several environmental variables (table 15), in combinations of 4 to 5 variables each, produced significant models. Only the top 4- and 5-variable solutions are shown in table 15. The highest overall correlation (Rho=0.98, P=0.007, table 15) was attained with a 4-variable model that included the maximum pH in September, open canopy percent, the presence of local gravel bars, and bedrock. Because all algal samples were collected from riffles, some of the geomorphic influences on algae suggested by the BEST analyses could be governed through effects on the quality of riffle habitats, shading, or other factors, even through direct or indirect effects on benthic macroinvertebrates and grazing rates, although none of these mechanisms was evaluated during this study.

Among the water-quality variables, the afternoon (maximum) value of pH explained the greatest amount of variation in algal assemblages (table 15). Diatoms are especially sensitive to pH conditions, and pH is considered a fundamental factor in shaping the composition of diatom assemblages (Birks and others, 1990). Photosynthesis by algae also raises the pH through a shift in the inorganic carbon equilibrium (Wetzel, 1983) that results from algal uptake of carbon dioxide, so it is not surprising that the multivariate analysis would identify pH as being an important explanatory variable in shaping the diatom assemblages during summer. The high abundance of alkaliphilic diatoms in the river is also consistent with the alkaline pH values observed during summer, but it is unclear to what degree, if any, the alkaline pH affects the diatom assemblages, especially given the high similarity among samples.

The percentage of open canopy, as determined by Cole (2004) was important in the 4- and 5-variable BEST models (table 15). Light can have a strong effect on algae (Lyford and Gregory, 1990), and daily cycles of oxygen production are predictably suppressed, for example, by cloudy weather, which reduces the sunlight required for photosynthesis. In the Molalla River, light availability is also partly controlled by channel width and sinuosity, which were highest in the alluvial GR3 reach.

An active stream channel, indicated by a high channel sinuosity (table 15; fig. 15), widened active channels (fig. 16), and high channel migration rates (fig. 17) can increase light availability for algae by reducing the amount of riparian vegetation or the effective shade through increases in the channel width as observed at the Goods Bridge site near the downstream end of GR3. Then again, a highly mobile channel can facilitate reworking of gravels and cobbles during high flow events that may scour algae from streambed substrates, reduce biomass, and 'reset' the system. Alternatively, reworking of gravels during bankfull flow events may also produce a positive effect on the algal biomass by reducing benthic invertebrate grazers. For example, studies over an 18-year period on the Eel River in northern California found that scouring high flows exceeding bankfull discharge had profound effects on vulnerable benthic macroinvertebrates that echoed through the river food web up to fish (Powers and others, 2008). In the absence of bankfull flows during winter, high survival of algal grazers resulted in higher abundance of macroinvertebrates in spring, especially *Dicosmoecus* caddisflies that graze on filaments of *Cladophora* (Wooton and others, 1996). In contrast, winters with flows exceeding bankfull discharge reduced the invertebrate grazers, which resulted in greater abundance of algae (Powers and others, 2008). Flows in the Molalla River during the winter preceding the 2010 summer sampling peaked at about 240 m^3/s (8,500 ft^3/s), or well below bankfull discharge value for the 2-year recurrence interval of 382 m^3/s (13,500 ft^3/s). This may explain the high abundances of *Dicosmoecus* observed in the Molalla River during 2010, and why algal biomass levels were not exceptionally high. Although similar interactions between high flow events, benthic invertebrate populations, and algal assemblages observed in the Eel River also may occur in the Molalla River, deciphering such effects in future studies is complicated by the year to year variability in streamflow conditions and the tendency for the channel to transport bedload material.

In addition to grazers, another factor governing algal biomass is the length of the algal growing season, which in 2010 was truncated by the cool and cloudy weather. Algal biomass levels presented here may, therefore, be lower than what might occur in years with more sunshine. This fact emphasizes the importance of long-term data and documentation of conditions over a range in conditions so that a better understanding of factors controlling algal-invertebrate-fish dynamics can be developed.

Table 14.   Physical characteristics for five sites sampled for water quality and algae in the Molalla River, Oregon.

[**Abbreviations:** D_{10}, size of sediment for which 10 percent of the sample is finer; D_{50}, median particle size; D_{90}, size of sediment for which 90 percent of the sample is finer; FPkm, flood plain kilometer; km, kilometer; km^2, square kilometer; LiDAR, light detection and ranging; m, meter; mm, millimeter; m/m, meter per meter; m/s, meter per second; m/yr, meter per year; m^2/m, square meter per meter; m^3/s, cubic meter per second; n/a, no data available]

Variable	Molalla River sampling site				
	Upstream of Glen Avon Bridge	Upstream of Highway 211 Bridge	Upstream of Highway 213 Bridge	Upstream of Goods Bridge	Downstream of Knights Bridge
Instream/site characteristics					
	Dominant land cover				
	Forested	Forested/ rural/ agricultural	Rural/ agricultural/ forested	Rural/ agricultural	Rural/ agricultural/ urban
Elevation (m)	163	97	72	33	23
LiDAR river stationing (km)	44.20	31.07	23.40	10.50	4.75
D_{10} particle size (mm)	32	38	29	25	14
D_{50} particle size (mm)	108	71	95	71	41
D_{90} particle size (mm)	316	157	156	139	77
Sand, percent	1.25	0.0	0	2	5.5
Drainage area (km^2)	344	531	546	842	893
Low flow discharge (10 percent exceedance) (m^3/s)	0.90	1.39	1.43	2.21	2.34
Median discharge (50 percent exceedance) (m^3/s)	7.72	11.9	12.3	18.9	20.0
Bankfull discharge (2-year flood) (m^3/s)	157	243	250	385	408
Water velocity at low flow, 90 percent exceedance (m/s)	0.39	0.52	0.45	0.47	0.54
Water velocity at median discharge, 50 percent exceedance (m/s)	0.83	1.06	1.00	0.91	1.04
Water velocity at bankfull discharge (2-year flood) (m/s)	2.41	2.56	2.27	2.53	2.79
Averaged residual pool depth (m)	n/a	0.4	1.7	0.7	0.4
Average water surface slope for 1 kilometer reach (m/m)	0.005	0.004	0.003	0.002	0.001
Presence of local streambed bedrock	Yes	Yes	Yes	No	Yes
Presence of local gravel bars	No	Yes	Yes	Yes	No
Flood-plain and channel characteristics					
Geomorphic flood plain FPkm segment (km)	36	25	20	10	4
FPkm segment average flood plain width (km)	0.21	1.13	0.90	0.83	2.66
Local flood plain width, (km)	0.10	0.98	0.98	0.70	2.25
1994 active channel width (m)	29.5	38.8	44.7	44.1	30.3
2000 active channel width (m)	27.1	47.1	60.8	52.3	34.5
2005 active channel width (m)	27.4	35.1	39.8	47.5	35.0
2009 active channel width (m)	31.2	33.9	45.9	56.7	35.7
Averaged active channel width (m)	28.8	38.7	47.8	50.1	33.9
1994–2000 annual migration rate (m/yr)	1.02	2.27	3.43	1.62	1.11
2000–2005 annual migration rate (m/yr)	1.00	0.55	0.88	1.15	0.48
2005–2009 annual migration rate (m/yr)	1.36	1.58	1.92	3.70	2.05
Time-weighted annual migration rate (m/yr)	1.10	1.52	2.19	2.00	1.14
Gemorphic reach average percent confined banks	44.9	21.1	21.1	15.1	24.2
FPkm segment average percent confined banks	23.4	48.1	26.0	41.2	51.3
Local bank confinement (number: 0,1,2 banks confined)	0	2	0	2	2
1994 channel sinuosity	1.10	1.35	1.35	1.29	1.05
2000 channel sinuosity	1.12	1.39	1.39	1.45	1.08
2005 channel sinuosity	1.10	1.39	1.39	1.44	1.08
2009 channel sinuosity	1.10	1.40	1.40	1.29	1.09
Average sinuosity	1.10	1.38	1.38	1.37	1.07
2009 specific bar area in FPkm segment (m^2/m)	6.68	24.73	56.60	19.47	8.31
2009 percentage of bare vegetation on bars	55.15	83.36	37.96	30.85	74.43
2009 percentage of moderate vegetation on bars	0	0	56.66	69.15	14.47
2009 percentage of dense vegetation on bars	44.85	16.64	5.38	0	11.10

Table 15. Summary of BEST variables and top models explaining patterns in diatom species relative abundance in the Molalla River, Oregon, July and August 2010.

[Algal taxa relative abundance values for July and August were averaged for this analysis. **Abbreviation:** D_{50}, median particle size; mm, millimeter]

Variable	Variable / model correlation and significance	
	Rho	P=
Individual variables		
Presence of local gravel bars	0.85	
Maximum pH in September	0.84	
2009 channel sinuosity	0.84	
2009 specific bar area	0.71	
Averaged active channel width	0.52	
D_{50} particle size (mm)	0.39	
Open canopy percent	0.35	
Flood plain width (segment)	0.27	
Low flow discharge (10 percent exceedance)	0.09	
Presence of local streambed bedrock	-0.07	
Top combination models		
Top 4-Variable Model	0.98	0.007
Presence of local gravel bars		
Maximum pH in September		
Open canopy percent		
Presence of local streambed bedrock		
Top 5-Variable Model	0.98	0.012
Presence of local gravel bars		
2009 channel sinuosity		
D_{50} particle size (mm)		
Open canopy percent		
Presence of local streambed bedrock		

The two other geomorphic variables identified in the BEST analyses were the presence of local gravel bars and bedrock, which can influence algal assemblages through various mechanisms and that operate on different spatial and temporal scales. Cobble and gravel bars provide ideal habitat for algae to colonize and grow, while bedrock can direct groundwater to the surface through features such as contact springs that may deliver nutrients that fuel algal growth.

Gravel bars in the alluvial section of the middle lower Molalla River can serve many potential ecosystem functions. The gravels are generally clean with little sediment, which provides high quality spawning habitat, and water flows into and out of gravel bars. Such hyporheic flows may produce cool water patches as they do in other rivers (Ebersole and Liss, 2003), and flow through the hyporheic zones or associated habitats can lower nutrients through various biogeochemical processes (Davis and others, 2011; Bencala, 2005; Mulholland and DeAngelis, 2000). Gravel bars also work to aerate surface water, and in the Molalla River, minimum dissolved oxygen levels were 84–87 percent of saturation in August and September 2010, which was sufficient to support a host of benthic organisms for riverine food webs. Benthic respiration in cobble and gravel habitats could contribute to offsetting increases in pH from algal photosynthesis or bring pH down, as was observed in this study in the reach from Glen Avon Bridge to Highway 211. The fact that afternoon pH values now occasionally approach and sometimes reach the State standard of 8.5 during summer places importance on sustaining ecosystem function of such gravel habitats by limiting introduction of fine sediments or "fining" of the coarse productive cobbles. At times, benthic respiration and production of CO_2 in the alluvial GR3 reach appear to more than offset the decrease in CO_2 associated with photosynthesis. In August and September 2000, for example, a 0.5 unit decline in afternoon pH was observed between Highway 213 and Goods Bridge (fig. 26C and D). This may have resulted from greater respiration by benthic communities and (or) input of lower pH ground water or exchanges of hyporheic flows, or from less periphyton that year. Powell (1995) measured distinct declines in stream pH downstream of alluvial reaches of the Little River (North Umpqua Basin, Oregon) and increases in pH in open bedrock channels where algae was abundant but gravels few. In 2010, however, the alluvial reach of the Molalla River contained appreciable periphyton, and as a result there were modest increases in pH of 0.2–0.3 units downstream at Goods Bridge (fig. 26A and B) indicating that photosynthesis utilization of inorganic carbon outpaced respiration. The vast expanses of gravel bars and shallow riffles in the reach upstream of Goods Bridge also result in lowering nutrient concentrations for further algal growth in downstream reaches. Reductions in concentrations of soluble reactive phosphorus and ammonium were observed at the Goods Bridge site in both 2000 and 2010 (fig. 28). Nitrate concentrations, however, were higher at Goods Bridge and increased further downstream at Knights Bridge. This could have resulted from a higher net balance between uptake by algae and inputs of nitrate from tributaries such as Milk and Gribble Creeks, agricultural irrigation return flows, groundwater seeps, urban runoff, or other sources.

The presence of bedrock in a stream channel could influence algal assemblages in a number of ways. As mentioned previously, bedrock may bring groundwater to the surface within the channel or through contact springs along the banks, which could affect water chemistry by enhancing nutrient and major ion concentrations. Inputs of cooler groundwater could also, for example, moderate minimum and maximum water temperatures, as was observed in the nearby Clackamas River (Burkholder, 2007). Conversely, the absence of bedrock and dominance by alluvium in much of the reach upstream of Goods Bridge (GR3) could produce a habitat that is conducive to streamflow gains and losses (hyporheic exchange) that could affect the physical and hydraulic habitat conditions and the downstream water chemistry. Hyporheic flow into and out of gravel bars was commonly observed in the Molalla River, especially in the reach downstream of Highway 213. The daily maximum water temperature at Goods Bridge, although higher than at the Highway 213 site, was not as high as the longitudinal trajectory might suggest (fig. 24). In 2001, a very warm, low flow-water year, water temperatures did not increase much in the reach between Highway 213 and Knights Bridge (fig. 25), which could be from hyporheic cooling or inputs of cooler groundwater near the downstream end of the alluvial reach. Although not measured in this study, it is possible that shallow or regional groundwater enters the river upstream of where bedrock again emerges to the surface near Canby. The longitudinal pattern in streamflow during June 2000 (fig. 24C) showed about a 7 percent loss in flow at Rkm 15.8 despite the input of Milk Creek. Then, a 20 percent gain in flow was observed downstream at Canby. Although part of this increase can be attributed to input of Gribble Creek, the measuring site was downstream of the Canby drinking water intake, so the increase is likely larger.

Aquatic Habitat and Water-Quality Conditions for Fish

The Molalla River, a once highly productive salmon stream, continues to support a healthy population of benthic macroinvertebrates that feed on abundant diatoms and other algae in the river, but fish populations struggle (Oregon Department of Fish and Wildlife and National Oceanic and Atmospheric Administration, 2010). Although food resources appear adequate for fish populations, the native winter steelhead and spring Chinook salmon are challenged by several potential stressors including high water temperatures, introduced warm-water fish species, and contaminants such as pesticides that enter the lower river from tributaries draining agricultural and urban areas. Additionally, moderately high pH and low dissolved oxygen results from cycles of algal photosynthesis and benthic respiration that also may stress fish at times.

The physical habitat in the lower river includes deep holding pools for fish throughout much of the river corridor (fig. 10), but shallow-water habitat in many areas may affect salmon migration or native trout populations by contributing to elevated water temperatures that are unfavorable for fish, and by reducing available habitat. High-quality riffles suitable for salmon spawning are abundant and mostly free of fine sediment in areas exposed to flow, but increases in sand and silt in the lower river do appear to degrade riffle habitat to some degree.

Water temperatures continue to be a potentially limiting factor for fish populations, with maximum water temperatures of 22–24°C from Highway 211 Bridge to Knights Bridge in 2010. The steady warming in the study reach is caused, in part, by inputs of warm tributaries in the middle and lower reaches. Data collected by ODEQ for the temperature TMDL (Williams and Bloom, 2008) found even higher water temperatures in 2004, ranging between 24 and 26°C throughout the lower 40-km reach. TIR data collected for the TMDL found notable declines in water temperatures associated with inputs of colder tributaries, springs, seeps, and groundwater. Such areas are key habitats for fish and other aquatic life as they provide cooler "thermal refugia" such as holding pools during warm periods, and may be potential sites for future study, protection, or restoration.

Warm water can also hamper salmonid populations by favoring warm water fish species, promoting growth of disease-causing organisms, and lowering concentrations of dissolved oxygen. Low streamflows in summer, partly a natural phenomenon due to geologic factors (lack of porous High Cascades rocks in the basin), also contributes to producing high water temperatures and may also reduce dissolved oxygen levels. Concentrations of dissolved oxygen were lowest in the early morning, with two sites—Goods and Knights Bridges—having values less than 8 mg/L in the water column. Dissolved oxygen concentrations in the gravels of fish-spawning beds in the lower river are typically about 3 mg/L lower (Carter, 2005) and could affect successful incubation of salmonid eggs. Although low inter-gravel dissolved oxygen (IGDO) concentrations in the lower Molalla River might be an issue for nonnative fall Chinook salmon, the native runs of winter steelhead and spring Chinook salmon do not use the lower river for spawning (Oregon Department of Fish and Wildlife, 1992). Nevertheless, resident salmonids in the lower river could be affected by low levels of dissolved oxygen, so future studies could characterize IGDO concentrations to determine if this is affecting native fish populations.

Another potential water-quality limitation for fish populations and other aquatic life in the Molalla River is high pH. Although the pH is affected by geochemical factors such as the weathering of rocks, in productive rivers such as the Molalla, pH is strongly controlled by the amount of photosynthesis, which removes inorganic carbon, mostly as CO_2, from the water column. Photosynthesis results in a distinct diel increase in pH during the day, and respiration by all aquatic life reintroduces CO_2 causing pH to decline at night, reaching a minimum just before sunrise. Physiological effects of highly alkaline pH levels on fish may include reduced ability to excrete ammonia or regulate ion balance, and effect that is manifested within fish gills (Laurent and others, 2000). High pH can affect such physiologic processes; however, not much is known about the potential effects from a constantly changing pH. Although the pH values were within the 6.5–8.5 unit range, maximum values reached 8.4 units at Knights Bridge in September 2010 and instantaneous measurements by ODEQ were as high as 8.5 units in the lower river during past summers (fig. 27). Also, because pH and other field measurements (DO and water temperature) are typically taken in the main flow, such values may not characterize what salmonid fry, for example, experience given that they tend to congregate along stream margins, out of the main current where hydraulic flushing is less and algal abundance tends to be higher. Future studies, described below, could include continuous measurement of field parameters, or cross sectional measurements including along shorelines to better characterize the potential for negative effects on fish eggs, fry, or juveniles.

Implications for Resource Management

Multiple river-management options are available to resource managers tasked with addressing geomorphic, aquatic habitat, and water-quality issues in the Molalla River. These options differ in complexity, cost, and impact. More importantly, the efficacy of different strategies is strongly dependent on the geomorphic nature of the reach of interest. For example, one strategy may be best applied to GR3 while a separate strategy is best applied to GR5. Although it is outside the purview of the USGS to recommend specific river-management options, restoration engineers and scientists could use this report to guide design and implementation of future action plans (Runyon and Stout, 2011).

Recent overbank flooding, channel migration, and channel avulsions in select reaches have caused problems for property owners living along the Molalla River flood plain. The geomorphic assessment described herein showed that high flows, including the 1996 flood, have a strong effect over river channel processes, including channel migration rates, and on riparian vegetation that impact stream conditions. Less shade and channel widening cause warming in the river that promotes the growth of bacteria, fish parasites, and disease, which is exacerbated by the low streamflow and high temperatures in summer. The average annual low daily mean streamflow at Canby is just 1.7 m^3/s (60 ft^3/s), and daily mean streamflow has been as low as 0.6 m^3/s (22 ft^3/s in 1959). Because the watershed lacks high-elevation snowpack and there are no large reservoirs for water storage in the basin, summer low flows pose significant risks to fish and other aquatic life.

Looking into the future, it is unclear whether high flows would increase or decrease in severity. It has been suggested that high flows have increased in intensity as warmer global temperatures have more efficiently converted ocean evaporation into rainfall (Min and others, 2011). Along the west coast of Washington State and British Columbia, Mass and others (2011) showed increasing trends in extreme precipitation and associated peak flows from the 1950s to the present. In the same study, however, peak flow trends in the Coastal Range of Oregon were shown to be decreasing over the same period (Mass and others, 2011). A closer investigation of data from 11 streamflow-gaging stations in the Umpqua River basin in the Oregon Coast Range by Wallick and others (2010b) showed a general decreasing trend in peak flow magnitude over most of the 20th century. Regardless of the peak-flow trends along the Molalla River in the coming decades, it is reasonable to anticipate events rivaling the magnitudes of the 1964 and 1996 floods.

Changes in land use, both along the river corridor and in the upper catchment, may have a stronger impact on river processes than changes in climate. For example, the building of additional houses in the flood plain and the channel-migration zone could lead to increased confinement of the river corridor, as well as higher risk to the structures from natural river processes. Planners evaluating new structures or revetments in the river corridor should consider potential adverse effects on flooding potential, water quality, or aquatic habitat. Similarly, there is an opportunity with modern forestry practices in the upper catchment to minimize the input of sediment into the river corridor and perhaps reduce the flashiness of high flows in the Molalla River.

This study found moderately elevated levels of algal productivity and diel fluctuations in dissolved oxygen and pH symptomatic of excessive photosynthesis, but pH values were compliant with water-quality standards. Dissolved oxygen concentrations, however, could approach levels that are harmful for fish or developing fish eggs and larvae in years when algal abundance or stream temperatures are higher. Because standards are based on 30-day minimum concentrations, evaluation of the data collected for this study against the DO standard was not possible.

Another potential implication for river and fish management is that annual cycles of algal growth and accrual of biomass that provide a base to the food chain are highly variable from year to year depending on weather conditions, flows during colonization, and many other factors. Several years of monitoring over a range in conditions may be necessary to characterize the dominant processes well enough to inform conceptual and mathematical models for prediction purposes.

In 2010, the Molalla River had only moderate algal biomass and the potential for substantial grazing by herbivorous macroinvertebrates was clearly evident. However, studies in the Eel River in northern California demonstrated that benthic macroinvertebrates may be reduced by large flow events, leading to subsequent increases in algal biomass (Wooton and others, 1996; Powers and others, 2008). Given the potential importance of these processes in the Molalla River, future monitoring and focused studies could provide insights into river food webs, fish health, and productivity. Monitoring and experimental research could help develop new policies, guide adaptive management strategies, and establish realistic expectations for salmon recovery efforts according to hydrologic conditions. Such an understanding may also increase our ability to predict fish returns or anticipate effects of climate change and changes in streamflow conditions.

Options available for improving water quality in the Molalla River include targeted riparian tree planting to increase shade and reduce erosion of soils, bank naturalization (grading and stabilizing banks with vegetation to reduce erosion), and nutrient management to curtail non-point sources along the river and within the flood plain. Improving water-quality conditions by increasing flow in the river is challenged by the lack of persistent snowpack or large water storage reservoirs in the basin, but water conservation strategies might increase instream flows. Water allocations for instream flows are junior to most other uses, and withdrawals of water from a large number of wells in the basin for agricultural and domestic purposes may intercept flow to the river, potentially exacerbating the negative effects of low flows during summer.

Future management actions in the basin, including implementation of the recent TMDL for temperature, are anticipated to lower water temperatures and may result in other improvements to water quality over time. Restoration efforts organized by the Molalla River Improvement District, watershed councils, and other citizen groups including Molalla RiverWatch and the Molalla River Alliance may also contribute to improving habitat and water-quality conditions for fish and other aquatic life.

Issues with flooding in GR3 and elsewhere could be addressed by giving the river a wider corridor to flow within. Removal of levees in GR4 and GR5 could, for example, reconnect the river to overbank riparian areas and might help reduce downstream flood peaks by temporarily storing water during large floods.

Potential Future Studies

Future monitoring and focused studies could contribute to better understanding of river health in many ways. Implementation of a monitoring program to evaluate the effectiveness of possible changes in land or water management (flood-plain restoration projects, forestry activities, or water-conservation strategies, for example) might help evaluate and fine tune adaptive-management strategies to improve water-quality conditions.

Future studies also could more fully evaluate the potential for algae to cause high pH or low DO in the river, should conditions deteriorate over time. If this were to occur, nutrient source identification and reduction strategies could be initiated to curb excessive algal production. Basinwide synoptic sampling (see Carpenter, 2003) for nutrients and streamflow in the main-stem Molalla River and tributaries could identify the primary source areas and this information could be useful in developing best management strategies to reduce inputs.

Although field parameter data collected for the current study helped to characterize water-quality conditions, including daily minimum and maximum values for water temperature, pH, and dissolved oxygen, deployment of a continuous water-quality monitor in the lower Molalla River could provide indications of deteriorating water-quality. Such monitoring could better define the health of the Molalla River through time and analyses of the resulting data could alert managers to potential trends in water quality over time. Continuous data on field parameters also would complement ODEQ's ongoing ambient water-quality monitoring program, which samples the lower Molalla River several times a year.

Measurements of water temperature, pH, and dissolved oxygen along shorelines in areas of abundant algal growth could characterize the potential negative effects on fish eggs, fry, or juveniles due to adverse water-quality conditions. In addition, sampling of inter-gravel dissolved oxygen concentrations could determine whether ample dissolved oxygen is present in spawning riffles when eggs are incubating. Given that dissolved oxygen levels in redds are typically about 3 mg/L lower than DO levels in the water column, fish egg development could be adversely affected at the lowest DO concentrations observed in the Molalla River (slightly less than 8 mg/L).

Future studies also could examine and quantify algal photosynthesis rates, benthic respiration, and look for key groundwater/surface-water interaction zones in the river. Recent TIR data collected for ODEQ's temperature TMDL could be used to identify areas where groundwater and (or) hyporheic water emerge. Such studies could provide insights to the importance of alluvial habitats in providing cold water refugia, for example, but could also determine influxes of nutrients that fuel algal growths. The depth of alluvial fill in GR3 could be determined with geophysical approaches or by examining drilling logs for wells in the flood plain. The alluvial section of river, GR3, currently supports abundant benthic macroinvertebrates and appears highly productive, so periodic monitoring of benthic community and riffle habitats for sedimentation and embeddedness, for example, could provide warning signs should conditions change.

Another potentially strong influence on algae populations are benthic invertebrate grazers, which on the basis of high densities, appeared to keep algal biomass levels in check. Although this effect was not investigated or quantified during this study, additional studies could examine the effect of invertebrate grazing on algal biomass and species composition in the Molalla River through enclosure/exclosure mesocosm studies that vary the abundance and (or) composition of grazers, predatory macroinvertebrates, and fish (salmonids and introduced warm water fish), similar to experiments conducted in the Eel River by Powers and others (2008). These multiple year studies highlight the importance of long-term datasets and underscore the need to shift from a single species to whole-community ecosystem perspective when considering management options or developing action plans geared toward fish recovery.

Although nutrient levels in the Molalla River were not very high, uptake by periphyton can mask inputs of nutrient-rich groundwater that may feed algae from below. There is some evidence of water gains in the lower river that may be due to groundwater, and the cooler temperatures in the lower reaches of Gribble Creek also suggest groundwater inputs to the river in this area. Given the agricultural land use in the southwestern part of the Molalla River basin, such inputs could contain elevated concentrations of the nutrients nitrate and orthophosphate, or even agricultural pesticides. Nitrate concentrations were much higher at Knights Bridge than at the Goods Bridge site during both August and September samplings (fig. 28), which could be due to inputs (at Knights Bridge) from groundwater and other sources. Fine-scale seepage studies could be conducted to confirm and quantify these water inputs, and determine nutrient loading from these sources. Given the low nutrient concentrations observed, this might be a logical starting point for identifying nutrient sources if high algal biomass becomes an issue in the future, if drinking water quality deteriorates, or if pH and dissolved oxygen concentrations become unhealthy for fish and other aquatic life.

A full accounting of revetments through the river corridor is an important data set not yet assembled. These data could be used in the planning of restoration projects. A more detailed mapping of the geologic structure of the lower Molalla River, including the identification of geologic units, would aid the understanding of river profile evolution and groundwater/surface-water interactions. Hydraulic and sediment-transport modeling of the river corridor under current and future climate conditions would help planners anticipate shifts to the hydrologic and sediment-transport regime and resulting impacts on fish and people. New bioenergetics models couple river hydraulics with biological factors to evaluate spatially the quality of fish habitat. Applying these models to the Molalla River could provide important insight for fish biologists trying to understand why salmonids do not use the river system as much as other river systems in western Oregon. Finally, thermal models of the river system under future climate-change scenarios would help identify thermal stressors to the aquatic habitat that may be anticipated in the coming decades.

Acknowledgments

This work builds on previous efforts by many other scientists, including Michael Cole (ABR, Inc.), who conducted earlier studies on habitat, water quality, and benthic macroinvertebrates in the Molalla River. We thank Michael for his helpful insights and for providing datasets used in some of the analyses. The authors especially wish to acknowledge the late Robert "Bob" Reynolds, former President of the Molalla River Improvement District (MRID) for his vision, commitment to the river, and for his contribution and support of science. Alan Gallagher and Kathryn Duthies (MRID) and Kay Patteson (Molalla RiverWatch) were instrumental in bringing together various partners and provided data and other information that were used to develop the study. Within the USGS, Krista Jones, Joseph Mangano, and Xavier Rodriquez Lloveras collected pebble-count data throughout the study area, and Adam Stonewall completed the gaging-station analysis; Rose Wallick and James O'Connor provided geomorphic and geologic insights; Tana Haluska digitized and established the geomorphic flood plain; Matt Johnston, Amy Brooks, Karl Lee (retired USGS) and Megan Bela (USGS volunteer) collected water temperature and streamflow data in 2001. Special thanks to John English (Oregon Department of Geology and Minerals) and Sheri Schneider (USGS) for providing LiDAR data, and to the University of Oregon Libraries for providing historic aerial images. Marc Howatt (Public Works Director, City of Canby) and Brian Hutchins (Canby drinking-water plant facility manager) provided insights about water quality of the Molalla River and Pat Curran and Curt McLeod (Curran-McLeod, Inc.) provided some key hydrological insights. We also thank the many private landowners in the basin that provided access to sampling sites and valuable information on the history and condition of the Molalla River.

References Cited

Anderson, C.W., and Carpenter, K.D., 1998, Water quality and algal conditions in the North Umpqua River basin, Oregon, 1992–95, and implications for resource management: U.S. Geological Survey Water-Resources Investigations Report 98-4125, 78 p.

Aquatic Analysts, 2007, Algae analytical and quality assurance procedures—Lab methods, including sample preparation, microscopic identifications, data processing, and archiving: Aquatic Analysts, 4 p.

Bahls, L.L., 1993, Periphyton bioassessment methods for Montana streams: Helena, Mont., Water Quality Bureau, Department of Health and Environmental Sciences, 69 p.

Bahls, L.L., 2007, *Cymbella janischii,* giant endemic diatom of the Pacific Northwest—Morphology, ecology and distribution compared to *Cymbella mexicana*: Northwest Science, v. 81, no. 4, p. 284-292.

Baker, V.R., 1973, Paleohydrology and sedimentology of Lake Missoula flooding in eastern Washington: Geological Society of America Special Paper 144, 79 p.

Beechie, T.J., Liermann, M., Pollock, M.M., Baker, S. and Davies, J., 2006, Channel pattern and river-floodplain dynamics in forested mountain river systems: Geomorphology, v. 78, no. 1-2, p. 124-141.

Bencala, K.E., 2005, Hyporheic exchange flows, *in* Anderson, M., ed., Encyclopedia of hydrological sciences: John Wiley and Sons, Ltd., v. 3, pt. 10, chap. 113, p. 1733-1740.

Birks, H.J.B., Line, J.M., Juggins, S., Stevenson, A.C., and ter Braak, C.J.F., 1990, Diatoms and pH reconstruction: Philosophical Transactions of the Royal Society of London, v. 327, no. 1240, p. 263-278.

Boyd, M., and Kasper, B., 2003, Analytical methods for dynamic open channel heat and mass transfer—Methodology for the Heat Source model, version 7.0: Portland, Oreg., Watershed Sciences.

Bretz, J.H., 1925, The Spokane flood beyond the Channeled Scablands: Journal of Geology, v. 33, p. 97-115 and 236-259.

Bureau of Land Management and U.S. Forest Service, 1999, Molalla watershed analysis: Salem, Oreg., Bureau of Land Management, Salem District Office, 242 p., accessed December 20, 2011, at http://www.blm.gov/or/districts/salem/plans/files/watershed_analyses/sdo_molalla_wa/sdo_molalla_wa99.pdf.

Burkholder, B.K., 2007, Influence of hyporheic flow and geomorphology on temperature of a large, gravel-bed river, Clackamas River, Oregon, USA: Corvallis, Oregon State University, Master's thesis, 179 p.

Carpenter, K.D., 2003, Water-quality and algal conditions in the Clackamas River basin, Oregon, and their relations to land and water management: U.S. Geological Survey Water-Resources Investigations Report 2002-4189, 114 p. (Also available at http://pubs.usgs.gov/wri/WRI02-4189/.)

Carrick, H.J., Lowe, R.L., and Rotenberry, J.T., 1988, Guilds of benthic algae along nutrient gradients— Relationships to algal community diversity: Journal of the North American Benthological Society, v. 7, no. 2, p. 117–128.

Carter, K., 2005, The effects of dissolved oxygen on steelhead trout, coho salmon, and Chinook salmon-Biology and function by life stage: California Regional Water Quality Control Board North Coast Region report, 9 p.

Cazaubon, A., Rolland, T., and Loudiki, M., 1995, Heterogeneity of periphyton in French Mediterranean rivers: Hydrobiologia, v. 300/301, p. 105-114.

Church, M., 1988, Floods in cold climates, *in* Baker, V.R., Kochel, R.C., and Patton, P.C., eds., Flood geomorphology: New York, John Wiley and Sons, p. 205-229.

Clark, P.U., and Bartlein, P.J., 1995, Correlation of late Pleistocene glaciation in the western United States with North Atlantic Heinrich events: Geology, v. 23, p. 483-486.

Clarke, K.R. and Gorley, R.N., 2006, PRIMER v. 6, User Manual: Primer-E, Plymouth, United Kingdom, 190 p.

Cole, M.B., 2002, Assessment of macroinvertebrate communities of the Molalla River, Oregon: Final report by ABR, Inc., Environmental Research and Services prepared for Molalla RiverWatch, 24 p.

Cole, M.B., Blaha, R.J., and Killian, M.P., 2004, Lower Molalla River and Milk Creek watershed assessment: Final report by ABR, Inc., Environmental Research and Services prepared for Molalla RiverWatch, 114 p.

Conlon, T.D., Wozniak, K.C., Woodcock, D., Herrera, N.B., Fisher, B.J., Morgan, D.S., Lee, K.K., and Hinkle, S.R., 2005, Ground-water hydrology of the Willamette basin, Oregon: U.S. Geological Survey Scientific Investigations Report 2005-5168, 83 p.

Cooper, R.M., 2005, Estimation of peak discharges for rural, unregulated streams in western Oregon: U.S. Geological Survey Scientific Investigations Report 2005-5116, 134 p.

Cude, C., 1996, Oregon water quality index report for Middle Willamette basin, water years 1986–1995: Oregon Department of Environmental Quality, accessed July 8, 2009, at http://www.deq.state.or.us/lab/wqm/wqindex/midwill3.htm.

Cuffney, T.F., Meador, M.R., Porter, S.D. and Gurtz, M.E., 1997, Distribution of fish, benthic invertebrate, and algal communities in relation to physical and chemical conditions, Yakima River Basin, Washington, 1990: U.S. Geological Survey Water Resources Investigations Report 96-4280, 94 p.

Davis, J.H., Griffith, S.M., and Wigington, P.J., Jr., 2011, Surface water and groundwater nitrogen dynamics in a well drained riparian forest within a poorly drained agricultural landscape: Journal of Environmental Quality, v. 40, p. 505-516.

Dodds, W.K., 1993, What controls levels of dissolved phosphate and ammonium in surface waters?: Aquatic Sciences v. 55, no. 2, p. 132-142.

Ebersole, J.L., and Liss, W.J., 2003, Cold water patches in warm streams—Physicochemical characteristics and the influence of shading: Journal of the American Water Resources Association, v. 39, p. 355-368.

Gannett, M.W., and Caldwell, R.R., 1998, Geologic framework of the Willamette lowland aquifer system, Oregon and Washington: U.S. Geological Survey Professional Paper 1424-A, 32. p.

Garcia, M.H., 2008, Sediment transport and morphodynamics, *in* Garcia, M.H., ed., Sedimentation engineering—ASCE Manuals and Reports on Engineering Practice No. 110: Reston, Va., American Society of Civil Engineers, p. 21-164.

Hampton, E.R., 1972, Geology and ground water of the Molalla-Salem slope area, northern Willamette Valley, Oregon: U.S. Geological Survey Water-Supply Paper 1997, 83 p.

Hanks, T.C., and Webb, R.H., 2006, Effects of tributary debris on the longitudinal profile of the Colorado River in Grand Canyon: Journal Geophysical Research, v. 111, 13 p., doi:10.1029/2004JF000257.

Heyn, K., and Bassett, R., 2009, The ecological and recreational benefits of the Molalla River, Oregon: American Rivers and the Native Fish Society, 13 p.

Hillebrand, H., and Sommer, U., 1999, The nutrient stoichiometry of benthic microalgal growth—Redfield proportions are optimal: Limnology and Oceanography, v. 44, no. 2, p. 440-446.

Hubbard, L.L., 1991, Oregon floods and droughts, *in* National Water Summary 1988–89: U.S. Geological Survey Water-Supply Paper 2375, p. 459-466.

Jefferson, A., Grant, G., and Rose, T., 2006, Influence of volcanic history on groundwater patterns on the west slope of the Oregon High Cascades: Water Resources Research, v. 42, no. W12411, 15 p., doi:10.1029/2005WR004812.

Jefferson, A., Grant, G.E., and Lewis, S.L., 2004, A river runs underneath it—Geological control of spring and channel systems and management implications, Cascade Range, Oregon—Advancing the fundamental sciences: Proceedings of the Forest Service National Earth Science Conference, San Diego, Calif., October 18-22, p. 391-400.

Jones, J.L., 2006, Side channel mapping and fish habitat suitability analysis using LIDAR topography and orthophotography: Photogrammetric Engineering and Remote Sensing, v. 71, no. 11, p. 1202-1206.

Klingeman, P.C., 1973, Indications of streambed degradation in the Willamette Valley: Corvallis, Oregon State University, Department of Civil Engineering, Water Resources Research Institute Report WRRI-21, p. 99.

Knighton, D., 1998, Fluvial forms and processes: New York, Oxford University Press, Inc., 383 p.

Lamberti, G.A., Gregory, S.V., Hawkins, C.P., Wildman, R.C., Ashkenas, L.R., and Denicola, D.M., 1992, Plant-herbivore interactions in streams near Mount St. Helens: Freshwater Biology, v. 27, p. 237-247.

Laurent, P., Wilkie, M.P., Chevalier, C., Wood, C.M., 2000, The effect of highly alkaline water (pH 9.5) on the morphology and morphometry of chloride cells and pavement cells in the gills of the freshwater rainbow trout—Relationship to ionic transport and ammonia excretion: Canadian Journal of Zoology, v. 78, p. 307-319.

Likens, G.E. and Bilby, R.E., 1982, Development, maintenance and role of organic detritus dams in New England streams, *in* Swanson, F.J., Janda, R.A., Dunne, T., and Swanston, D.N., eds., Sediment budgets and routing in forested drainage basins: U.S. Forest Service Technical Report PNW-141, p. 122–128.

Lisle, T.E., 1987, Using 'residual depths' to monitor pool depths independently of discharge: U.S. Forest Service Research Note PSW-394, 4 p.

Lowe, R.L., and Pan, Y., 1996, Benthic algal communities as biological monitors, *in* Stevenson, J.R., Bothwell, M.L., and Lowe, R.L., eds., Algal ecology—Fresh-water benthic ecosystems: San Diego, Calif., Academic Press, Inc., p. 705–739.

Lyford, J.H., and Gregory, S.V., 1995, The dynamics and structure of periphyton communities in three Cascade Mountain streams: Proceedings of the International Association of Theoretical and Applied Limnology, v. 19, p. 1610–1616.

Macklin, J.H., 1948, Concept of the Graded River: Geological Society of America Bulletin, v. 59, p. 463–512.

Madej, M.A., 1999, Temporal and spatial variability in thalweg profiles of a gravel-bed river: Earth Surface Processes and Landforms, v. 24, p. 1153–1169.

Madej, M.A., and Ozaki, V., 2009, Persistence of effects of high sediment loading in a salmon-bearing river, northern California, *in* James, L.A., Rathbun, S.L., and Whittecar, G.R., eds., Management and restoration of fluvial systems with broad historical changes and human impacts: Boulder, Colo., Geological Society of America Special Paper 451, p. 43–55, doi: 10.1130/2008.2451(03).

Mass, C., Skalenakis, A., and Warner, M., 2011, Extreme precipitation over the west coast of North America—Is there a trend?: Journal of Hydrometeorology, v. 12, no. 2, p. 310–318, doi:10.1175/2010JHM1341.1.

McIntosh, B.A., Clarke, S.E., and Sedell, J.R., 1990, Bureau of Fisheries stream habitat surveys—Willamette River basin, summary report 1934-1942: Pacific Northwest Research Station, USDA Forest Service, Oregon State University, U.S. Department of Energy, Bonneville Power Administration, Division of Fish and Wildlife, Project No. 89-104, Contract No. DE-AI79-89BP02246, 492 electronic pages (BPA Report DOE/BP-02246-3).

Min, S.-K., Zhang, X., Zwiers, F.W., and Hegerl, G.C., 2011, Human contribution to more-intense precipitation extremes: Nature, v. 470, no. 7334, p. 378–381, doi:10.1038/nature09763.

Montgomery, D.R., and Buffington, J.M., 1997, Channel-reach morphology in mountain drainage basins: Geological Society of America Bulletin, v. 109, no. 5, p. 596–611.

Moulton, S.R., Kennen, J.G., Goldstein, R.M., and Hambrook, J.A., 2002, Revised protocols for sampling algal, invertebrate, and fish communities as part of the National Water-Quality Assessment Program: U.S. Geological Survey Open-File Report 02-150, 75 p.

Mulholland, P.J. and DeAngelis, D.L., 2000, Surface-subsurface exchange and nutrient spiraling, *in* Jones, J.B., and Mulholland, P.J., eds., Streams and ground waters: Academic Press, 149 p.

Mulholland, P.J., and Rosemond, A.D., 1992, Periphyton response to longitudinal nutrient depletion in a woodland stream—Evidence of upstream-downstream linkage: Journal of the North American Benthological Society, v. 11, no. 4, p. 405–419.

National Atmospheric and Oceanic Administration Fisheries, 2005, Endangered and Threatened Species—Final Listing Determinations for 16 ESUs of West Coast Salmon, and Final 4(d) Protective Regulations for Threatened Salmonid ESUs: Federal Register, v. 70, no. 123, p. 37,160–3,720,450, at http://www.nwr.noaa.gov/ESA-Salmon-Listings/Salmon-Populations/Chinook/CKUWR.cfm.

National Atmospheric and Oceanic Administration Fisheries, 2006, Endangered and Threatened Species-Final Listing Determinations for 10 Distinct Population Segments of West Coast Steelhead: Federal Register, v. 71, no. 3, p. 834–862, at http://www.nmfs.noaa.gov/pr/pdfs/fr/fr71-834.pdf

O'Connor, J.E., Jones, M.A., and Haluska, T.L., 2003, Flood plain and channel dynamics of the Quinault and Queets Rivers, Washington, USA: Geomorphology, v. 51, no. 1–2, p. 31–59.

O'Connor, J.E., Sarna-Wojcicki, A.M., Wozniak, K.C., Polette, D.J., and Fleck, R.J., 2001, Origin, extent, and thickness of Quaternary geologic units in the Willamette Valley, Oregon: U.S. Geological Survey Professional Paper 1620, 52 p.

O'Connor, J.E., and Benito, G., 2009, Late Pleistocene Missoula Floods—15,000–20,000 calendar years before present from radiocarbon dating: 2009 Portland Geological Society of America Annual Meeting, Portland, Oreg., October 18-21, Abstracts with Programs, v. 41, no. 7, p. 169, Paper No. 55-7.

Oregon Department of Environmental Quality, 1988, Oregon assessment of non-point sources of water pollution, cited in Bureau of Land Management and U.S. Forest Service (1999).

Oregon Department of Environmental Quality, 2003, Fish use designations, Willamette Basin, Oregon: Oregon Department of Environmental Quality, accessed December 20, 2011, at http://www.deq.state.or.us/wq/rules/div041/fufigures/figure340a.pdf.

Oregon Department of Environmental Quality, 2005, Designated salmon and steelhead spawning use designations, Willamette Basin, Oregon: Oregon Department of Environmental Quality, accessed December 20, 2011, at http://www.deq.state.or.us/wq/rules/div041/fufigures/figure340b.pdf.

Oregon Department of Environmental Quality, 2008, Oregon water-quality index summary report water years 1998–2007: Report 09-LAB-008, at Oregon Department of Environmental Quality, 2010, accessed January 9, 2011, at http://www.deq.state.or.us/lab/wqm/docs/09-LAB-008.pdf

Oregon Department of Environmental Quality, 2010, LASAR web application: Oregon Department of Environmental Quality database, accessed February 2, 2011, at http://deq12.deq.state.or.us/lasar2/.

Oregon Department of Fish and Wildlife, 1992, Molalla and Pudding subbasin fish management plan: Oregon Department of Fish and Wildlife, 110 p.

Oregon Department of Fish and Wildlife and National Oceanic and Atmospheric Administration Fisheries, 2010, Willamette River fish recovery-upper Willamette River conservation and recovery plan for Chinook salmon and steelhead: Public review draft, 29 p., accessed October 2010 at http://www.dfw.state.or.us/fish/CRP/docs/upper_willamette/UWR%20FRN2%20Mainbody%20final.pdf.

Oregon Department of Geology and Mineral Industries, 2010, LiDAR Collection and Mapping: Oregon Department of Geology and Mineral Industries, Oregon LiDAR Consortium (OLC), accessed December 20, 2011, at http://www.oregongeology.org/sub/projects/olc/default.htm.

Oregon Geospatial Enterprise Office, 2010, Oregon Digital Orthophoto Quads (DOQs): State of Oregon database, accessed December 20, 2011, at http://gis.oregon.gov/DAS/EISPD/GEO/data/doq.shtml.

Peterson, D.A., Porter, S.D., and Kinsey, S.M., 2001, Chemical and biological indicators of nutrient enrichment in the Yellowstone River basin, Montana and Wyoming, August 2000—Study design and preliminary results: U.S. Geological Survey Water-Resources Investigations Report 01-4238, 6 p. (Also available at http://pubs.usgs.gov/wri/wri014238/.)

Porter, S.D., 2008, Algal attributes—An autecological classification of algal taxa collected by the National Water-Quality Assessment Program: U.S. Geological Survey Data Series 329, 18 p. (Also available at http://pubs.usgs.gov/ds/ds329/.)

Powell, M.A., 1995, Report on pH in the Jackson Creek and Little River drainage basins of the Umpqua National Forest: Roseburg, Oreg., U.S. Department of Agriculture, Forest Service, Umpqua National Forest, Draft report to the U.S. Forest Service, [variously paged].

Powers, M.E., Parker, M.S., and Dietrich, B.E., 2008, Seasonal reassembly of a river food web—Floods, droughts, and impacts on fish: Ecological Monographs, v. 78, no. 2, p. 263–282.

Redfield, A.C., 1934, On the proportions of organic derivatives in sea water and their relation to the composition of plankton, *in* Daniel, R.J., ed., James Johnstone Memorial Volume: University Press of Liverpool, p. 177–192.

Redfield, A.C., 1958, The biological control of chemical factors in the environment: American Scientist, v. 46, no. 3, p. 205–221.

Rinella, F.A., and Janet, M.L., 1998, Seasonal and spatial variability of nutrients and pesticides in streams of the Willamette Basin, Oregon—1993–95, U.S. Geological Survey Water-Resources Investigations Report 97-4082-C, 59 p.

Runyon, J.R. and Stout, T., 2011, Middle and lower Molalla River restoration action plan: Prepared for Molalla River Watch and Molalla River Improvement District by Cascade Environmental Group, LLC, Portland, Oreg., 60 p.

Scholz, N L., Truelove, N., French, B.L., Berejikian, B.A., Quinn, T.P. Casillas, E. and Collier, T.K., 2000, Diazinon disrupts anti-predator and homing behaviors in Chinook salmon (*Oncorhynchus tshawytscha*): Canadian Journal of Fisheries and Aquatic Sciences, v. 57, p. 1911–1918.

Smelser, M.G., and Schmidt, J.C., 1998, An assessment methodology for determining historical changes in mountain streams: Fort Collins, Colo., U.S. Department of Agriculture, U.S. Forest Service Rocky Mountain Research Station, General Technical Report RMRS-GTR-6, 29 p.

Soil Survey Staff, 2008, Spatial and tabular data of the Soil Survey for Clackamas County, Oregon: U.S. Department of Agriculture, Natural Resources Conservation Service, accessed 2010 at: http://soildatamart nrcs.usda.gov/.

Spaulding, S.A., and Elwell, L., 2007, Increase in nuisance blooms and geographic expansion of the freshwater diatom *Didymosphenia geminata*: U.S. Geological Survey Open-File Report 2007-1425, 38 p.

Stevenson, R.J., Bothwell, M.L., and Lowe, R.L., 1996, Algal ecology—Freshwater benthic ecosystems: San Diego, Calif., Academic Press, Inc., p. 321–340.

Stock, J.D., and Montgomery, D.R., 1999, Geologic constraints on bedrock river incision using the stream power law: Journal of Geophysical Research, v. 104, no. B3, p. 4983–4993.

Swanson, F.J. and Lienkaemper, G.W., 1978, Physical consequences of large organic debris in Pacific Northwest streams: U.S. Department of Agriculture, Forest Service, Pacific Northwest Forest and Range Experiment Station, General Technical Report PNW-69, 12 p.

Taylor, G.H. and Hatton, R.R., 1999, The Oregon weather book, a state of extremes: Corvallis, Oreg. State University Press, 224 p.

Tetra Tech and KCM, Inc., 2000, City of Molalla wastewater facilities plan: Prepared for the City of Molalla, 168 p.

U.S. Department of Agriculture, 2010, Imagery Programs—NAIP imagery:Geospatial Data Gateway: U.S. Department of Agriculture database: accessed May 2010 at http:// datagateway nrcs.usda.gov.

U.S. Geological Survey, 1981 (rev.), Guidelines for determining flood flow frequency: U.S. Geological Survey Bulletin 17B of the Hydrology Subcommittee, Interagency Advisory Committee on Water Data, accessed December 20, 2011, at http://water.usgs.gov/osw/bulletin17b/ bulletin_17B html.

U.S. Geological Survey, 2006, Collection of water samples (ver. 2.0): U.S. Geological Survey Techniques of Water-Resources Investigations, book 9, chap. A4, September 2006, accessed January 9, 2011, at http://pubs.water.usgs. gov/twri9A4/.

U.S. Geological Survey, 2011, Clackamas River water quality monitors: U.S. Geological Survey database, accessed December 21, 2011, at http://or.water.usgs.gov/clackamas/ monitors/.

Vannote, R.L., Minshall, G.W., Cummins, K.W., Sedell, J.R., and Cushing, C.E., 1980, The river continuum concept: Journal of Canadian Fisheries and Aquatic Sciences, v. 37, p. 130–137.

Waite, I.R., Sobieszczyk, S., Carpenter, K.D., Arnsberg, A.J., Johnson, H.M., Hughes, C.A., Sarantou, M.J., and Rinella, F.A., 2008, Effects of urbanization on stream ecosystems in the Willamette River basin and surrounding area, Oregon and Washington: U.S. Geological Survey Scientific Investigations Report 2006-5101-D, 62 p.

Wallick, J.R., Anderson, S.W., Cannon, C., and O'Connor, J.E., 2010a, Channel change and bed-material transport in the lower Chetco River, Oregon: U.S. Geological Survey Scientific Investigations Report 2010-5065, 68 p.

Wallick, J.R., O'Connor, J.E., Anderson, S.W., Keith, M., Cannon, C., and Risley, J.C., 2010b, Channel change and bed-material transport in the Umpqua River basin, Oregon: U.S. Geological Survey Open-File Report 2010-1314, 135 p.

Walton, S.P., Welch, E.B., and Horner, R.R., 1995, Stream periphyton response to grazing and changes in phosphorus concentration: Hydrobiologia, v. 302, p. 31-46.

Watershed Sciences, 2009a, LiDAR remote sensing data collection: Department of Geology and Mineral Industries, Oregon Department of Forestry, Puget Sound LiDAR Consortium: Prepared for Puget Sound Regional Council, Kitsap County, Oregon Department of Geology and Mineral Industries and Oregon Department of Forestry, 131 p.

Watershed Sciences, 2009b, LiDAR remote sensing data collection: Department of Geology and Mineral Industries, Willamette Valley Phase I Oregon: Prepared for Oregon Department of Geology and Mineral Industries, 43 p.

Wehr, A., and Lohr, U., 1999, Airborne laser scanning— An introduction and overview: ISPRS Journal of Photogrammetry and Remote Sensing 54, p. 68–82.

Welch, E.B., Jacoby, J.M., Horner, R.R., and Seeley, M.R., 1988, Nuisance biomass levels of periphytic algae in streams: Hydrobiologia, v. 157, p. 161–168.

Wetzel, R.G., 1983, Limnology, 2nd ed.: Philadelphia, Saunders College Publishing, 858 p.

Wilde, F.D., Radtke, D.B., Gibs, Jacob, and Iwatsubo, R.T., eds., 2004 with updates through 2009, Processing of water samples (ver. 2.2): U.S. Geological Survey Techniques of Water-Resources Investigations, book 9, chap. A5, April 2004, accessed January 9, 2011, at http://pubs.water.usgs.gov/twri9A5/.

Williams, K.F., and Bloom, J., 2008, Molalla-Pudding Subbasin total maximum daily load (TMDL) and Water Quality Management Plan (WQMP): Oregon Department of Environmental Quality, accessed December 20, 2011, at http://www.deq.state.or.us/wq/tmdls/docs/willamettebasin/MolallaPudding/MoPudExecutiveSummary.pdf.

Wolman, M.G., 1954, A method of sampling coarse river-bed material: Transactions of the American Geophysical Union, v. 35, no. 6, p. 951–956.

Wooton, J.T., Parker, M.S., and Power, M.E., 1996, Effects of disturbance on river food webs: Science, v. 273, no. 5281, p. 1558–1561.

Wustman, B.A., Gretz, M.R., and Hoagland, K.D., 1997, Extracellular matrix assembly in diatoms (Bacillariophyceae)—A model of adhesives based on chemical characterization and localization of polysaccharides from the marine diatom Achnanthes longipes and other diatoms: Journal of Plant Physiology, v. 113, p. 1059–1069.

Appendixes and Evaluation of Quality Assurance Data

Evaluation of Quality Assurance Data

Quality assurance data tables are presented in the appendices and the results are summarized below. Two equipment blank samples were collected for nutrients, resulting in no detections for the first sample and two low-level detections in the second sample: 0.1 mg/L total Kjeldahl nitrogen (TKN) and 0.014 mg/L soluble reactive phosphorus (appendix D). The TKN detection was at the laboratory reporting limit and represents a low-level detection for this constituent. The ambient stream concentrations of TKN were generally low, about 2–3 times higher than the concentration reported in the blank. The phosphorus detection in the one blank sample was higher than any concentration measured in the stream; the detection in the blank, therefore, appears to be an isolated case. There were two replicate nutrient samples (appendix D) which, for the most part, were within ±10 percent. Exceptions occurred for dissolved ammonia, which had relatively high percent relative differences of 69 and 90 percent, but were within reason given the low concentrations as all replicate values were less than the laboratory reporting level of 0.02 mg/L. The percent relative differences for the two TKN replicate samples ranged from 6 to 20 percent, but were reasonably close considering that all values were relatively low and close to the laboratory reporting level of 0.1 mg/L.

One set of replicate samples analyzed from separate collections of stream cobbles were analyzed for algal biomass, species identification and enumeration. Although the replicate biomass samples were reasonably close (within 7 mg chlorophyll-a/m^2), these samples differed considerably with respect to the species composition (appendix E). About one-half of the species were identified in only one of the two replicates. Based on the relative density, the top three dominant taxa were the same for both samples, but based on biovolume, one sample (replicate #2) was dominated by the large stalked diatom *Gomphoneis herculeana* whereas this taxa was not identified in the first replicate sample (appendix E). This difference highlights the often patchy growth of periphyton (Cazaubon and others, 1995), particularly for macroalgae, which in the Pacific Northwest includes *G. herculeana* and other large stalked diatoms (Bahls, 2007), filamentous greens (*Cladophora* and others), and filamentous blue-green algae including *Oscillatoria*, *Phormidium*, and others (Anderson and Carpenter, 1998; Carpenter, 2003). These types of algae are much larger than typical diatoms and other microalgae and may be missed under high magnification, especially in 100 algal unit counts used for this study. Abundance values from the two replicate samples were averaged for all further analyses.

Appendixes A-G are included in a Microsoft© Excel workbook, which can be downloaded from http://pubs.usgs.gov/sir/2012/5017/.

Appendix A.   Particle size distributions collected using Wolman (1954) Pebble Counts on nine freshly deposited bars along the river corridor of the Molalla River, Oregon, September–October 2010.

Appendix B.   Periphyton species abundance data (cell density) in the Molalla River, Oregon, August–September 2010.

Appendix C.   Periphyton species abundance data (algal biovolume) in the Molalla River, Oregon, August–September 2010.

Appendix D.   Quality assurance data for total and dissolved nutrients in the Molalla River, Oregon, August–September 2010.

Appendix E.   Quality assurance data for replicate periphyton species composition samples from the Molalla River, Oregon, 2010.

Appendix F.   Nutrient and field parameter data for the Molalla River, Oregon, June 26, 2000.

Appendix G.   Field parameter data for the Molalla River, Oregon, 2000.